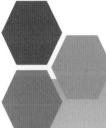

Introduction to Renewable Energy

RENEWABLE ENERGIES SERIES

skills2learn
www.skills2learn.com
Experts in e-learning & virtual reality simulation

CENGAGE
Learning

Australia • Brazil • Japan • Korea • Mexico • Singapore • Spain • United Kingdom • United States

**Introduction to Renewable Energy,
1st Edition
Skills2Learn**

Publishing Director: Linden Harris

Commissioning Editor: Lucy Mills

Development Editor: Lauren Darby

Senior Project Editor: Alison Burt

Senior Manufacturing Buyer: Eyvett Davis

Typesetter: MPS Limited

Cover design: HCT Creative

For product information and technology assistance,
contact **emea.info@cengage.com.**

For permission to use material from this text or product,
and for permission queries,
email **emea.permissions@cengage.com.**

British Library Cataloguing-in-Publication Data

A catalogue record for this book is available from the British Library.

ISBN: 978-1-4080-6465-8

Cengage Learning EMEA

Cheriton House, North Way, Andover, Hampshire, SP10 5BE
United Kingdom

Cengage Learning products are represented in Canada by Nelson Education Ltd.

For your lifelong learning solutions, visit **www.cengage.co.uk**

Purchase your next print book, e-book or e-chapter at **www.cengagebrain.com**

Printed in China by RR Donnelley
1 2 3 4 5 6 7 8 9 10 – 15 14 13

Contents

Foreword

The energy sector is a significant part of the UK economy and a major employer of people. It has a huge impact on the environment and plays a massive role in our everyday life, shaping both our work and domestic habits and processes. With environmental issues such as climate change and sustainable sourcing of energy now playing an important role in our society, there is a need to educate a significant pool of people about future technologies, with renewable energies in all likelihood playing an increasingly significant part in our total energy requirements.

This environmental and renewable energy series of e-learning programmes and text workbooks has been developed to provide a structured blended learning approach that will enhance the learning experience, stimulate a deeper understanding of the renewable energy trades and give an awareness of sustainability issues. The content within these learning materials has been aligned as far as is currently possible to the units of the National Occupational Standards and can be used as a support tool whilst studying for any relevant vocational qualifications.

The uniqueness of this renewable energy series is that it aims to bridge the gap between classroom-based and practical-based learning. The workbooks provide classroom-based activities that can involve learners in discussions and research tasks as well as providing them with understanding and knowledge of the subject. The e-learning programmes take the subject further, with high quality images, animations and audio further enhancing the content and showing information in a different light. In addition, the e-practical side of the e-learning places the learner in a virtual environment where they can move around freely, interact with objects and use the knowledge and skills they have gained from the workbook and e-learning to complete a set of tasks whilst in the comfort of a safe working environment.

The workbooks and e-learning programmes are designed to help learners continuously improve their skills and provide confidence and a sound knowledge base before getting their hands dirty in the real world.

About e-Consortia

This series of renewable energy workbooks and e-learning programmes has been developed by the e-Renewable Consortium. The consortium is a group of colleges and organizations that are passionate about the renewable energy industry and are determined to enhance the learning experiences of people within the different trades or those that are new to it.

The consortium members have many years experience in the renewable energy and educational sectors and have created this blended learning approach of interactive e-learning programmes and text workbooks to achieve the aim of:

- Providing accessible training in different areas of renewable energy
- Bridging the gap between classroom-based and practical-based learning
- Providing a concentrated set of improvement learning modules
- Enabling learners to gain new skills and qualifications more effectively
- Improving functional skills and awareness of sustainability issues within the industry
- Promoting health and safety in the industry
- Encouraging training and continuous professional development.

For more information about this renewable energy consortium please visit: **http://skills2learn.cengage.co.uk/9-renewable-energy**

About e-learning

INTRODUCTION

This renewable energies series of workbooks and e-learning programmes uses a blended learning approach to train learners about renewable energy skills. Blended learning allows training to be delivered through different mediums such as books, e-learning (computer-based training), practical workshops, and traditional classroom techniques. These training methods are designed to complement each other and work in tandem to achieve overall learning objectives and outcomes.

E-LEARNING

The Introduction to Renewable Energies e-learning programme that is also available to sit alongside this workbook offers a different method of learning. With technology playing an increasingly important part of everyday life, e-learning uses visually rich 2D and 3D graphics/animation, audio, video, text and interactive quizzes, to allow you to engage with the content and learn at your own pace and in your own time.

E-ASSESSMENT

Part of the e-learning programme is an e-assessment 'End test'. This facility allows you to be self-tested using interactive multimedia by answering questions on the e-learning modules you will have covered in the programme. The e-assessment provides feedback on both correctly and incorrectly answered questions. If answers are incorrect the learner is advised to revisit the learning materials they need to study further.

BENEFITS OF E-LEARNING

Diversity – E-Learning can be used for almost anything. With the correct approach any subject can be brought to life to provide an interactive training experience.

Technology – Advancements in computer technology now allow a wide range of spectacular and engaging e-learning to be delivered to a wider population.

Captivate and Motivate – Hold the learner's attention for longer with the use of high quality graphics, animation, sound and interactivity.

Safe Environment – E-Practical scenarios can create environments which simulate potentially harmful real-life situations or replicate a piece of dangerous equipment, therefore allowing the learner to train and gain experience and knowledge in a completely safe environment.

Instant Feedback – Learners can undertake training assessments which feedback results instantly. This can provide information on where they need to re-study or congratulate them on passing the assessment. Results and Certificates could also be printed for future records.

On-Demand – Can be accessed 24 hours a day, 7 days a week, 365 days of the year. You can access the content at any time and view it at your own pace.

Portable Solutions – Can be delivered via a CD, website or LMS. Learners no longer need to travel to all lectures, conferences, meetings or training days. This saves many man-hours in reduced travelling, cost of hotels and expenses amongst other things.

Reduction of Costs – Can be used to teach best practice processes on jobs which use large quantities of, or expensive materials. Learners can practise their techniques and boost their confidence to a high enough standard before being allowed near real materials.

INTRODUCTION TO RENEWABLE ENERGIES E-LEARNING

The aim of the renewable energies e-learning programme is to enhance a learner's knowledge and understanding of the renewable technologies. The course content is aligned to units from the Environmental National Occupational Standards (NOS) so can be used for study towards certification.

The programme gives the learners an understanding of the different types of heat and electricity producing technologies, the selection process plus the incentives available and the policies that apply to installations.

By using and completing this programme, it is expected that learners will:

- Explain what renewable energy is
- Explain how the main heat producing technologies work
- Explain how the main electricity producing technologies work
- List the incentives available and policies that apply to renewable energy installation
- Conduct an evaluation and audit to optimize the savings and costs associated with renewable energy installation.

The e-learning programme is divided into the following learning modules:

- Getting Started
- Introduction to Renewable Energy
- Heat Producing Technologies
- Electricity Producing Technologies
- Incentives and Policy
- Selection Process
- End Test

THE RENEWABLE ENERGIES SERIES

As part of the renewable energies series the following e-learning programmes and workbooks are available. For more information please visit: **http://skills2learn.cengage.co.uk/9-renewable-energy**

- Heat Pumps
- Solar Thermal Hot Water
- Solar PV
- Building Heat Loss Calculator (programme only)
- Solar Radiation Calculator (programme only)

About the NOS

The National Occupational Standards (NOS) provide a framework of information that outline the skills, knowledge and understanding required to carry out work-based activities within a given vocation. Each standard is divided into units that cover specific activities of that occupation. Employers, employees, teachers and learners can use these standards as an information, support and reference resource that will enable them to understand the skills and criteria required for good practice in the workplace.

The standards are used as a basis to develop many vocational qualifications in the United Kingdom for a wide range of occupations. This workbook and associated e-learning programme are aligned to the Environmental National Occupational Standards and the information within relates to the following units:

- Plan for Environmental Technology Systems, Equipment and Components
- Install Environmental Technology Systems, Equipment and Components
- Test Environmental Technology Systems, Equipment and Components
- Commission Environmental Technology Systems, Equipment and Components
- Inspect Environmental Technology Systems, Equipment and Components
- Diagnose Faults in Environmental Technology Systems, Equipment and Components
- Rectify Faults in Environmental Technology Systems, Equipment and Components
- Service and Maintain Environmental Technology Systems, Equipment and Components.

About the book

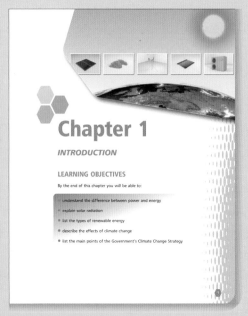

Learning Objectives at the start of each chapter explain the skills and knowledge you need to be proficient in and understand by the end of the chapter.

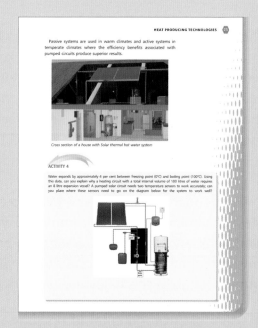

Activities are practical tasks that engage you in the subject and further your understanding.

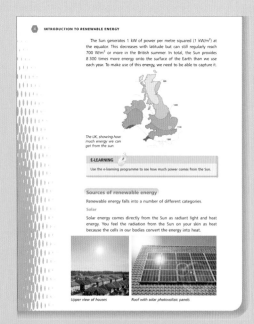

E-Learning Icons link the workbook content to the e-learning programme.

Sustainability Boxes provide information and helpful advice on how to work in a sustainable and environmentally friendly way.

Note on UK Standards draws your attention to relevant building regulations.

Functional Skills Icons highlight activities that develop and test your Maths, English and ICT key skills.

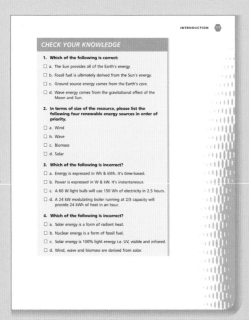

Check Your Knowledge at the end of each chapter to test your knowledge and understanding.

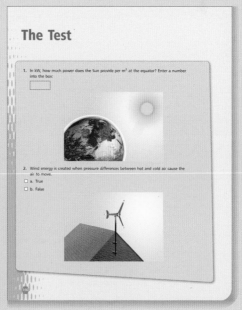

End Test in Chapter 6 checks your knowledge on all the information within the workbook.

Chapter 1

INTRODUCTION

LEARNING OBJECTIVES

By the end of this chapter you will be able to:

- understand the difference between power and energy

- explain solar radiation

- list the types of renewable energy

- describe the effects of climate change

- list the main points of the Government's Climate Change Strategy

inexhaustible In renewable energy source terms, inexhaustible sources are those which will not run out.

Effects of climate change

biomass Biological material which is derived from living, or recently living organisms, such as wood, waste (hydrogen) gas and alcohol fuels. It is commonly plant matter grown to generate electricity or produce heat. Biomass is a renewable energy source.

RENEWABLE ENERGY

Introduction

Renewable energy is energy that comes from an **inexhaustible** source – one that is infinitely replenished. Renewable energy comes from natural resources such as sunlight, wind and water, which are constantly replenished so they will never run out. Types of renewable energy include:

geothermal Earth's temperature increases with depth, outward heat flows from a hot interior.

- solar energy from radiation from the Sun

- wind energy from changes in atmospheric pressures

- hydro energy from the water in rivers, lakes and reservoirs

- **biomass** energy from the processing of organic matter

tidal Tides are the rise and fall of sea levels caused by the combined effects of the gravitational forces exerted by the Moon and the Sun and the rotation of the Earth.

- wave energy from the movement of the sea

- **geothermal** energy from the heat of the Earth's core

- **tidal** energy from the gravitational pull of the Moon and Sun.

Cross section of the Earth

Power and energy

Energy is the physical capacity for activity and power is the rate at which energy is generated or consumed. Power is measured in watts (W) and kilowatts (kW) and energy is measured in watt hours (Wh) and kilowatt hours (kWh). For example, an average low energy light bulb transforms 12 watts of electrical energy into light (and heat) every hour. The light bulb has a power rating of 12 watts and consumes 12 watt hours of energy in one hour. A typical one bar electric heater transforms one kilowatt of electrical energy into heat (and light) in one hour. The heater has a power rating of one kilowatt (1 kW) and consumes one kilowatt hour (1 kWh) of energy in one hour.

watts The watt is a derived unit of power in the International System of Units (SI). The unit measures the rate of energy conversion. It is defined as one joule per second.

kilowatts The kilowatt is equal to one thousand watts. This unit is used to express the output power of engines, the power consumption of tools and machines, the heating and cooling power used and any other forms of power.

kilowatt hours The kilowatt hour is a unit of energy equal to 1000 watt hours or 3.6 megajoules. The unit of energy is how much power is consumed in a time period e.g. kilowatt hours.

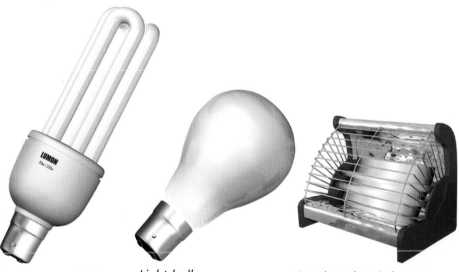

Low energy light bulb   *Light bulb*   *One bar electric heater*

The Sun generates 1 kW of power per metre squared (1 kW/m^2) at the equator. This decreases with latitude but can still regularly reach 700 W/m^2 or more in the British summer. In total, the Sun provides 8 300 times more energy onto the surface of the Earth than we use each year. To make use of this energy, we need to be able to capture it.

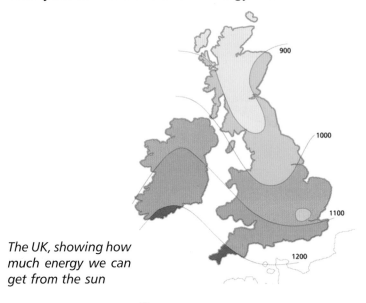

900

1000

1100

1200

The UK, showing how much energy we can get from the sun

E-LEARNING

Use the e-learning programme to see how much power comes from the Sun.

Sources of renewable energy

Renewable energy falls into a number of different categories.

Solar

Solar energy comes directly from the Sun as radiant light and heat energy. You feel the radiation from the Sun on your skin as heat because the cells in our bodies convert the energy into heat.

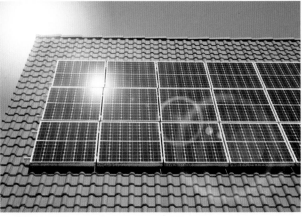

Upper view of houses   *Roof with solar photovoltaic panels*

Solar energy can be converted by photovoltaic panels into electricity or by solar thermal collectors to heat energy.

By using the deserts to generate solar power energy, we can obtain very large quantities of energy. However, this desert energy is expensive both to make and to transport.

It is important to remember that the majority of the energy used on Earth is derived from the Sun's light energy, as coal, oil and gas are in fact very long-term stored forms of solar energy, however they are not renewable. Only the energy from gravity (e.g. tidal) and the energy from the Earth (e.g. nuclear and geothermal) are not originally derived from solar energy.

> photovoltaic The word photovoltaic comes from photo which means light and voltaic which means electricity. Solar Photovoltaic or PV cells convert the energy from the Sun directly into electricity.

Wind

Wind energy is created when the Sun heats the atmosphere. Hot air rises and creates an area of low pressure whereas colder, denser air creates an area of high pressure as this air drops through the atmosphere. Where the pressure difference is great enough, air will move from the high pressure area to the low pressure area creating wind.

This is why hurricanes are formed in the tropics as the Sun's energy is much stronger there and so creates bigger storms with much more energy in the form of wind and rain. Wind energy is maintained over flat oceans which is why fairly small islands such as Britain and Ireland have plenty of wind energy.

Wind turbines

Hydro

Hydro energy ultimately comes from the Sun. Clouds are formed by evaporation of the Earth's water. They are moved by the wind over land and where these clouds are forced upwards by hills and mountains they cool and condense into rain. Movement of this water, either by the natural gravity of rivers and streams or the gravity of man-made dams, provides us with hydro energy. Therefore, hydro power uses both solar power and gravitational forces to create its energy.

Waterfalls are a source of hydro energy

Biomass

Biomass energy comes from organic matter which has used the Sun to photosynthesize and grow. There are numerous ways to process this energy including **anaerobic digestion** and burning wood.

anaerobic digestion Is a series of processes in which micro-organisms break down biodegradable material in the absence of oxygen.

photosynthesis Using the energy from sunlight, it is the process that converts carbon dioxide into organic compounds, especially sugars. Photosynthesis occurs in plants, algae and many species of bacteria.

SUSTAINABILITY

Biomass is a term used to describe all animal and vegetable organic matter. Everything that grows on Earth is fed directly from the Sun via **photosynthesis** or from eating from the food chain that starts with photosynthesis. There are many ways of processing this organic matter to make either food for human consumption or energy to improve quality of life. There is a lot of ongoing discussion about crops for food or for energy.

Grain collector tractor

Waves

Wave energy is created when the wind blows over the surface of the sea. Air pressure differences between the upwind and lee sides of a wave, as well as friction on the water surface, cause the waves to grow.

We create waves when we blow over a still body of bathwater. The same process on a much larger scale occurs in our oceans as the wind passes over the water's surface and engineers are now developing ingenious methods for capturing this wave energy.

Waves

Geothermal

Most geothermal energy is not derived from the Sun, instead it comes from the Earth's core. A small amount of geothermal energy, correctly called ground source energy, can be counted as ultimately coming from the Sun as the top 15m of the surface of the Earth are warmed by the Sun.

Earth's geothermal energy originates from the original formation of the planet (20 per cent) and from the radioactive decay of minerals (80 per cent). The geothermal gradient, which is the temperature difference between the Earth's central core and its surface, drives a continuous conduction of thermal energy in the form of heat from the core to the surface. This conduction varies in strength depending on the conductivity of the underlying rock and strata, and in areas where the conductivity is high, heat energy can be obtained for either generating electricity (high grade heat energy) or providing space heating (lower grade heat energy).

Ground source energy is mainly solar energy which heats the Earth's outer core. Some ground source energy comes from the Earth's core but it is just a small percentage of the overall ground source energy.

Geothermal energy comes from the Earth's core. Ground source energy comes from the Sun heating the surface of the Earth.

An example of geothermal energy

A geothermal power station

Tidal

Tidal energy is also not derived from the Sun's radiation. This energy is created by the gravitational pull of the Sun and the Moon on the seas and oceans.

Tides are the rise and fall of sea levels caused by the combined effects of the gravitational forces exerted by the Moon and the Sun and the rotation of the Earth. A tidal generator converts the energy of tidal flows into electricity. Greater tidal variation and higher tidal current velocities can dramatically increase the potential of a site for tidal electricity generation.

Tidal energy is derived from the gravitational pull of the Sun and Moon

Tidal energy generation

E-LEARNING

Use the e-learning programme to see more information about the different sources of renewable energy.

ACTIVITY 1

The following word search contains the seven renewable energies, three fossil (coal, oil, gas) and one other (nuclear). Can you find all eleven types of energy?

```
G   B   Y   B   J   X   P   L   Y   R   I
C   M   E   B   V   S   O   L   A   R   J
A   S   H   Y   D   R   O   F   B   B   M
C   J   T   T   P   Y   Y   F   I   A   F
R   L   A   M   R   E   H   T   O   E   G
J   A   C   M   T   O   F   M   M   S   Z
B   A   E   H   L   I   Y   J   A   E   C
S   E   N   L   V   K   D   Y   S   V   O
W   I   N   D   C   G   Y   A   S   A   A
N   E   O   I   L   U   A   Q   L   W   L
M   F   G   J   T   N   N   S   G   D   A
```

BIOMASS	OIL
COAL	SOLAR
GAS	TIDAL
GEOTHERMAL	WAVES
HYDRO	WIND
NUCLEAR	

CLIMATE CHANGE

What causes climate change?

More than 90 per cent of scientists agree that man-made climate change, specifically the rise in global temperature, is happening. Significant amounts of carbon dioxide are released into the atmosphere through burning fossil fuels for heating, generating electricity and travel. This causes the greenhouse effect because the carbon dioxide, along with other greenhouse gases, traps the heat energy from the Sun within the Earth's atmosphere.

The average global temperature has risen by around 0.75°C in the last 100 years. The Intergovernmental Panel on Climate Change has

climate change This is a change in the statistical distribution of weather over periods of time that range from decades to millions of years. Climate change may be limited to a specific region, or may occur across the whole Earth. Human activities such as burning fossil fuels which emit greenhouse gases, contribute to climate change. In the UK, 40 per cent of emissions are caused by individuals.

carbon dioxide (CO_2) A colourless, odourless gas that is formed during combustion, respiration and organic decomposition. Carbon dioxide emissions are considered to be a major cause of climate change.

fossil fuels These are fuels such as oil and coal that are formed by natural resources such as anaerobic decomposition of buried dead organisms. The age of the organisms and their resulting fossil fuels is typically millions of years, but can exceed two billion years. These fuels contain a high percentage of carbon and hydrocarbons.

greenhouse effect
This is a process by which reflected and infrared energy leaving a planetary surface is absorbed by some atmospheric gases, called greenhouse gases.

greenhouse gases
Gases in an atmosphere that absorb and emit radiation within the thermal infrared range. This process is the fundamental cause of the greenhouse effect.

Intergovernmental Panel on Climate Change (IPCC) IPCC is a scientific intergovernmental body tasked with evaluating the risk of climate change caused by human activity. The panel was established in 1988 by the World Meteorological Organization (WMO) and the United Nations Environment Programme (UNEP).

concluded that most of the temperature increase since the middle of the 20th century was most likely to have been caused by increasing concentrations of greenhouse gases arising from human activity such as fossil fuel burning and deforestation

Temperature increase is due to an increased concentration of greenhouse gases

How the greenhouse effect happens

As the Earth turns, the incoming solar radiation passes through the clear atmosphere.

Solar radiation passes through the atmosphere

Around 70 per cent of this radiation reaches the Earth's surface and this will be approximately 1 kW/m2 on a bright clear day at the equator when the Sun is directly overhead.

70 per cent of solar radiation reaches the Earth's surface

The rest of this solar radiation, approximately 30 per cent, is reflected by the Earth's atmosphere back into space.

30 per cent is reflected back into space

When that 70 per cent solar radiation reaches the Earth's surface, most of it is absorbed and then radiated back into the Earth's atmosphere as infrared heat.

Most of the 70 per cent of solar radiation reaches the earth surface

The greenhouse gases in the atmosphere, which are mainly composed of carbon dioxide with some methane and other gases, trap most of this heat and raise the temperature of the atmosphere. Without this effect, the heat would be lost into space and the Earth would be too cold to support any kind of life. However, in recent times, the increase in greenhouse gases, particularly the increase in carbon dioxide, is causing too much heat to be trapped and so causing the planet's overall temperature to rise. Because this process is caused by the products and by-products of human activity this is man-made climate change.

Greenhouse gases will trap most of the heat, raising the temperature of the atmosphere

To summarize, greenhouse gases trap much of the Sun's energy in the form of heat and raise the temperature of the Earth's atmosphere. Man-made climate change is caused by the increase in greenhouse gases. The right amount of greenhouse gases is essential to support life on planet Earth. However, the increase in greenhouse gases may have catastrophic consequences.

E-LEARNING

Use the e-learning programme to see a demonstration of the greenhouse effect.

What are the effects of climate change?

According to the International Panel on Climate Change (IPCC), the effects of climate change will be dramatic. Global warming is expected to be highest in the Arctic and would be associated with ice caps melting, causing sea levels to rise, which will have grave consequences for lowland areas.

SUSTAINABILITY

Our weather patterns will change dramatically with more extreme weather events and changes to the type and amount of food that can be grown. Some plant and animal species will become extinct and millions of people will be at risk. In the light of this, it makes sense to start to look for alternative renewable sources of energy.

We therefore need to find an alternative to burning fossil fuels.

Effects of climate change

ACTIVITY 2

Some of the Sun's energy is:
a. transmitted through the atmosphere to the Earth's surface
b. reflected back into space
c. absorbed into the Earth's surface
d. re-emitted as infrared radiation.

If some of the Sun's energy is not trapped in the Earth's atmosphere, no life would be supported. However, if too much energy is trapped, the planet will be too hot to support life as we know it today. On the diagram on page 14 mark where the Sun's energy is:

- transmitted

- reflected into space

- absorbed by the Earth's atmosphere

Continued on next page

- absorbed by the Earth's surface
- re-emitted by the Earth's surface
- lost into space
- infrared energy absorbed by the Earth's atmosphere.

Kyoto Protocol The Kyoto Protocol is a protocol to the United Nations Framework Convention on Climate Change (UNFCCC or FCCC), aimed at fighting global warming.

Bali Road Map After the 2007 United Nations Climate Change Conference on the island of Bali in Indonesia, the participating nations adopted the Bali Road Map as a two-year process to finalizing a binding agreement for 2009 in Copenhagen.

GOVERNMENT POLICIES

Global government strategies

UK AND INTERNATIONAL STANDARDS

The 1997 Kyoto Protocol is recognized as the most important piece of legislation relating to climate change in recent history. It came into effect in February 2005. In June 2007, 172 member states and countries ratified the agreement to set legally binding targets for the limitation or reduction of their greenhouse gas emissions.

In December 2007 what has become known as the Bali Road Map was formally adopted by the European Union and a number of other countries. The objective of the Road Map is to gain unilateral commitment to limit global warming to no more than 2°C above the pre-industrial temperature.

The Bali conference was followed by two meetings, the first in Poland in 2008 and then Copenhagen in 2009 where details of the binding national commitments were agreed in the Copenhagen Accord. The full Copenhagen Accord is available online.

The negotiations on the post-2012 climate protection regime were originally due to be concluded at the Copenhagen climate change conference in December 2009. However, following very difficult negotiations, the Copenhagen meeting only achieved a political agreement, the Copenhagen Accord, which lists some key elements of future climate protection policy. The Accord is not binding and was only taken note of by the meeting of the conference.

The aim of Germany and the EU to adopt a new comprehensive and legally binding post-2012 climate protection agreement was not achieved.

Nevertheless, the Copenhagen Accord was a step in the right direction. More than 140 countries (including the EU Member States) have declared their formal support for the Copenhagen Accord. Numerous industrialized and developing countries have submitted specific emission reduction targets for 2020.

It was also decided in Copenhagen that the negotiations in the two parallel working groups under the UN Framework Convention on Climate Change and the Kyoto Protocol should be continued on the basis of the existing negotiating texts until the next Climate Change Conference in Cancún, Mexico, which took place from 29 November to 10 December 2010. Despite difficult negotiations, a package of decisions was adopted at the end of the two week conference – the Cancún Agreements. These lay down the contents of the Copenhagen Accord in United Nations decisions and in some cases also go beyond this. For the first time, a UN decision recognized the 2°C target. The next climate change conference took place from 28 November to 9 December 2011 in Durban, South Africa. It is expected that these conference rounds will be ongoing for the foreseeable future as the world addresses climate change issues.

Copenhagen Accord
The Copenhagen Accord is the document that delegates at the United Nations Climate Change Conference (UNCCC) agreed to 'take note of' at the final plenary session of the Conference on 18 December 2009.

Kyoto Protocol came into effect in February 2005

UK government strategies

In this workbook, we are using the UK as a case study of a country that is in the middle of the climate change process. Some countries such as Austria, Denmark and Germany are world leaders on climate change issues and have created many jobs from this sector. Other countries are only just starting on their climate change journeys. The UK relied on natural gas from the North Sea for much of the 1990s and so started later than some other north European nations on tackling climate change issues but is still ahead of other countries that had developmental, political or other factors delaying the process.

The UK needs to secure clean, safe, affordable energy to heat and power its homes and businesses. Creating a low carbon and resource efficient world means making major structural changes to the way all of us work and live, including how we source, manage and use our energy.

Legally binding targets have been set for the UK and Europe:

- UK – to source 15 per cent renewable energy by 2020
- Europe – to source 20 per cent renewable energy by 2020

In the UK, a new government department, the **Department of Energy and Climate Change**, has been created to take the lead in tackling climate change with seven Departmental Strategic Objectives.

Every EU member state has its own legally binding target as set out in the Renewable Energy Sources (RES) Directive and its own starting point for the quantity of renewable energy needed in their overall energy supply mix.

Department of Energy and Climate Change (DECC) This is a British government department created on 3 October 2008 to take over some of the functions of the Department for Business, Enterprise and Regulatory Reform (energy) and Department for Environment, Food and Rural Affairs (climate change).

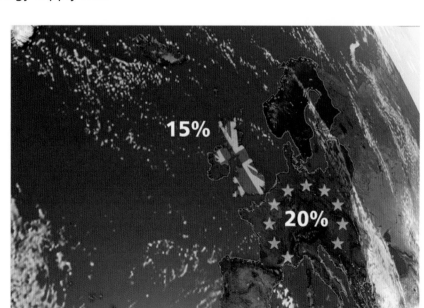

By 2020, the UK aims to source 15 per cent renewable energy and Europe 20 per cent

CHECK YOUR KNOWLEDGE

1. Which of the following is correct:

☐ a. The Sun provides all of the Earth's energy

☐ b. Fossil fuel is ultimately derived from the Sun's energy.

☐ c. Ground source energy comes from the Earth's core.

☐ d. Wave energy comes from the gravitational effect of the Moon and Sun.

2. In terms of size of the resource, please list the following four renewable energy sources in order of priority.

☐ a. Wind

☐ b. Wave

☐ c. Biomass

☐ d. Solar

3. Which of the following is incorrect?

☐ a. Energy is expressed in Wh & kWh. It's time-based.

☐ b. Power is expressed in W & kW. It's instantaneous.

☐ c. A 60 W light bulb will use 150 Wh of electricity in 2.5 hours.

☐ d. A 24 kW modulating boiler running at 2/3 capacity will provide 24 kWh of heat in an hour.

4. Which of the following is incorrect?

☐ a. Solar energy is a form of radiant heat.

☐ b. Nuclear energy is a form of fossil fuel.

☐ c. Solar energy is 100% light energy i.e. UV, visible and infrared.

☐ d. Wind, wave and biomass are derived from solar.

5. Which of the following is correct?

☐ a. Climate change is completely man-made.

☐ b. Climate change is completely natural i.e. man has no influence.

☐ c. Kyoto, Copenhagen, Bali & Durban have all held environmental conferences.

☐ d. IPCC stands for International Policy on Climate Change.

6. Which of the following is incorrect?

☐ a. The UK has to achieve 20 per cent renewable energy by 2020.

☐ b. Europe has to achieve 20 per cent renewable energy by 2020.

☐ c. Every EU member state has its own binding RE target for 2020.

☐ d. Countries such as Austria, Denmark and Germany have created many jobs from renewable energy.

7. Which two renewable energies are not produced by the Sun's radiation?

☐ a. Biomass

☐ b. Geothermal

☐ c. Solar

☐ d. Tidal

☐ e. Wave

☐ f. Wind

8. Which of the following statements about climate change are true?

☐ a. All radiation from the sun is absorbed by the earth.

☐ b. Greenhouse gases in the atmosphere trap heat and raise the temperature of the atmosphere.

☐ c. Millions are highly likely to be at risk of extreme weather events if measures are not taken against climate change.

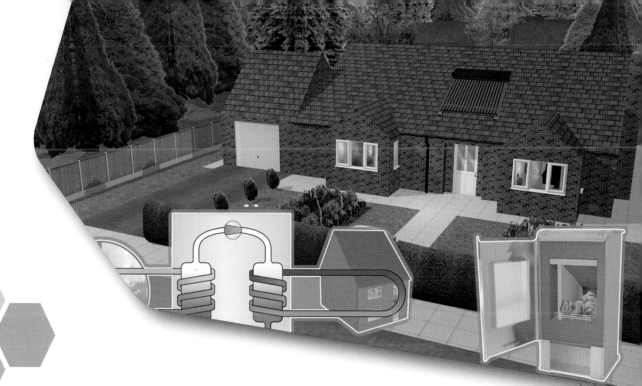

Chapter 2

HEAT PRODUCING TECHNOLOGIES

LEARNING OBJECTIVES

By the end of this chapter you will be able to:

- describe the different forms of heat producing technologies

- explain where solar thermal hot water would be applicable

- explain where ground source and air source heat pumps would be applicable

- explain why biomass fuel is carbon neutral and where it is applicable

- explain where micro combined heat and power would be applicable

- list the advantages and disadvantages of each heat producing technology and list the building location, features, fabric, regulations and planning permission that would apply to each technology

SOLAR THERMAL HOT WATER

Introduction

Despite the weather, the UK and most of Europe receives a reasonable source of solar energy. This resource increases somewhat across the Middle East and Africa. However, most climates can obtain plenty of solar energy; for example the UK receives about half the solar energy received in the sunniest areas. This can be captured and converted into heat using solar thermal water heating.

This energy can then be used to provide heat for **domestic hot water**, swimming pools and also, less frequently, space heating. In fact, solar water heating could provide up to 70 per cent of domestic hot water in the UK with most houses obtaining at least half of their hot water requirements from solar energy.

Before the installation of solar thermal water heating or any other form of renewable heating, it is important that the insulation and draught proofing of the building is made as efficient as possible. Energy efficiency is always the first priority on any cost saving project as reducing consumption saves more money than providing the energy from a renewable resource. It is also necessary to calculate the direction and angle of the roof on which the solar panels are to be mounted for optimum performance, and to consider the distance between the solar heating system and the distribution system.

> **domestic hot water (DHW)** Water which is heated and supplied for washing and bathing via taps or showers. DHW is not necessarily potable water and should be handled carefully to manage both Legionella and scalding risks.

Solar thermal water heating system

E-LEARNING

Use the e-learning programme to see a demonstration of Solar Thermal Heating.

Solar thermal hot water

Cross section of a house with a solar thermal hot water system

The main component of a solar thermal hot water system is the solar collector which captures the solar radiation, transfers it to the **heat transfer** fluid and then uses it to heat the water in the hot water cylinder. There are two main types of collector:

heat transfer The movement of heat from one medium to another.

- flat plates
- evacuated tubes.

Flat plate collectors provide a higher efficiency at a lower temperature difference. They can be integrated into the roof of a building and are generally cheaper than evacuated tubes. A typical house with four occupants would require 4m² of flat plate collector.

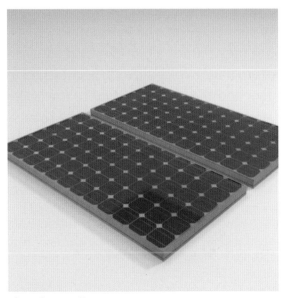

Flat plate collectors

Evacuated tubes consist of rows of parallel glass tubes in a vacuum to reduce convection and heat loss. They are excellent at reaching high temperatures and, for a given surface area, can be up to 20 per cent more efficient than flat plate collectors. However, they tend to be more expensive than flat plate collectors and cannot be integrated into the roof. A typical house with four occupants would require $3m^2$ to $3.5m^2$ of evacuated tube collector.

As the fuel source is free, efficiency is only one of many deciding factors. Flat plates or evacuated tubes might look more attractive or integrate better into different buildings and roof types. Also, if roof space is restricted, evacuated tubes can offer a lower surface area option.

Evacuated tubes

Ideally the existing hot water storage cylinder should be replaced with a larger volume twin-coil cylinder. This means the water can also be heated from the existing primary heating system. With a twin-coil system, the bottom coil is connected to the solar collector and the top coil is connected to the existing heat source such as a boiler or heat pump. During the winter season, the bottom half of the cylinder acts as a 'preheat' zone for the 'other heat source' zone, whilst during the sunny season, the whole cylinder can be left for the solar energy to do most, if not all, of the work.

An alternative is to have two cylinders; one attached to the solar thermal heating system and the other attached to the boiler or heat pump heating system. In this system arrangement, the solar system preheats all the water that also passes through the other heat source cylinder.

UK AND INTERNATIONAL STANDARDS

When replacing a hot water cylinder, both building and water regulations and standards must be followed. Indeed, this applies to all renewable energy interventions. Competent plumbers, roofers and electricians are required to ensure professional installations. Always check you have up-to-date information for the local area you are working in.

Hot water cylinders

Solar thermal heating systems can be designed to work with either vented or unvented cylinders or with thermal stores. In the diagram below, twin-coil systems are shown but please note that two cylinder systems can also be specified.

A vented system has a cold water storage tank that is open to the atmosphere. This means that as the water temperature rises and the water expands, the build up of pressure can be safely managed by expansion of the water in the cold water storage tank.

An unvented system is sealed and the temperature, volume and pressure in the system must be safely controlled by an expansion vessel and a series of other safety features.

Twin-coil systems

To summarize, the design of the system requires a choice between a flat plate or evacuated tube collector, either of which can be linked to a vented, unvented or thermal store cylinder system. Vented, unvented and thermal store systems can be specified in either twin-coil or two cylinder circuit layouts.

E-LEARNING

There are many other factors to consider in solar thermal system design that are covered in more detail in the Solar Thermal e-learning programme.

ACTIVITY 3

Using the following drawings as a template, draw a sketch layout of an evacuated tube collector linked to a twin-coil cylinder, and a flat plate collector linked to a two cylinder layout.

Active and passive solar circuits

heat exchanger A heat exchanger is a device designed to efficiently transfer heat from one medium to another.

Active systems require a pump to move the cooler water from the heat exchanger in the cylinder back to the solar collector and then, once heated, return the heated water back to the heat exchanger.

A passive system, which is also called a thermosyphon system, relies on the natural circulation which occurs when hot water rises and cooler water sinks, thereby creating the circulation within the solar circuit.

Thermosyphon systems are also sometimes called gravity fed systems because it's the forces of gravity that drive the heavier, colder water down as the lighter, hotter solar heated water travels upwards.

Passive systems are used in warm climates and active systems in temperate climates where the efficiency benefits associated with pumped circuits produce superior results.

Cross section of a house with Solar thermal hot water system

ACTIVITY 4

Water expands by approximately 4 per cent between freezing point (0°C) and boiling point (100°C). Using this data, can you explain why a heating circuit with a total internal volume of 100 litres of water requires an 8 litre expansion vessel? A pumped solar circuit needs two temperature sensors to work accurately; can you place where these sensors need to go on the diagram below for the system to work well?

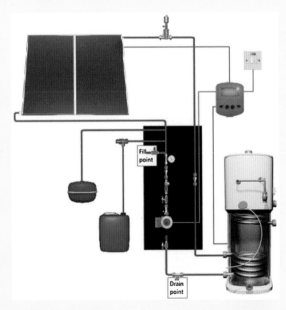

Solar thermal hot water system

building control certificate A certificate that will be issued by an approved inspector for the compliance with Building Regulations in the proposed building work plans.

Solar Trade Association An association representing UK-based solar companies. The association promotes the uptake of solar technologies to domestic, commercial and industrial applications.

Installing a solar thermal hot water system

Location, planning and technical standards

UK AND INTERNATIONAL STANDARDS

In England and Wales, planning permission is not required for roof or wall mounted solar panels provided they do not protrude beyond the existing line of the roof by more than 200mm. Exceptions to this are listed buildings, road-facing roofs in conservation areas and World Heritage sites. This planning permission process is similar in many countries; some states provide a precedent towards installing renewable energy systems and others require a detailed application process, so always check local requirements before installing any new systems.

Solar systems usually require a building control certificate, for further information, you should consult your local Solar Trade Association for the latest advice. Building control or building standards are used to make sure the necessary technical installation standards are followed. Planning permission is the process of making sure that the new installation matches the local planning and appearance requirements of the area in question.

Solar panels are remarkably flexible in where they can be fitted. Ideally a south-facing roof with a 20° to 50° pitch should be used. However, solar panels can be fitted to east or west facing roofs if the panel area is increased by 25 per cent to compensate for the loss in efficiency. Some key points to consider are to keep the length of pipework from the solar collector to the cylinder as short and well insulated as possible and to always use an accredited installer to do the work. Keeping the pipework between the solar collector and cylinder short ensures that the maximum amount of solar heat is collected and used, and using an accredited competent installer makes sure that a professional and safe job is completed.

Roof top view of a house with solar panels

ACTIVITY 5

You have been asked to survey a house with three bedrooms and four occupants located in the centre of England. The customer wants a solar water heating system and the roof has a pitch of 30° and is east-facing. The existing cylinder is a vented single coil with a volume of 150 litres. You have a recommended volume of 50 litres per occupant. You need to specify an appropriate solar system for this property.

- Specify the collector surface area if flat plates are used
- Specify the collector surface area if evacuated tubes are used
- State the first choice and size of solar cylinder
- State whether a pumped or thermosyphon solar circuit should be used

HEAT PUMPS

An overview of heat pumps

Heat pumps work by extracting low temperature heat from the environment. This heat can be stored in the ground, in water such as lakes, canals and rivers and in the air. The heat is drawn in from the environment by the heat pump, where it undergoes a compression and expansion cycle to increase the temperature.

The increased temperature is then passed to a distribution system such as underfloor heating, a hot water system or heating for a swimming pool. In this way, heat pumps 'upgrade' the heat from the colder source, i.e. the cooler outside air or ground temperature, and distribute this upgraded heat around a heating distribution circuit. A heat pump can also work in reverse to act as a cooling system.

turbines A turbine is a rotary engine that extracts energy from a fluid flow and converts it into useful work. Please note that a fluid can be a gas or a liquid.

ground source heat pumps (GSHP) A system that uses heat from the ground to heat (or cool) a building. It uses the Earth as a heat source in the winter or a heat sink in the summer.

Although heat pumps use electricity, they are classed as a source of renewable energy as they save on fossil fuels and the electricity to run them can come from a renewable source, for example, wind turbines.

It is worth noting that a heat pump will only save energy compared to a conventional heating system if the heat pump is well designed, installed, commissioned and maintained.

Good heat pump design ensures that the difference between the source temperature and the distribution temperature is as small as can be. This is why ground source heat pumps (GSHP) tend to be slightly more efficient than air source heat pumps (ASHP) as the ground tends to be warmer in winter, the main heating season. However, GSHPs, because of the ground works, tend to be more expensive to install, so the decision between the two systems is usually a fine balance.

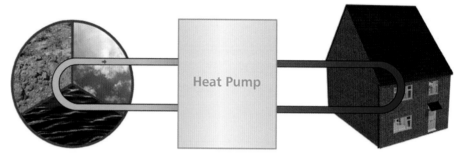

Simple illustration of a heat pump

How a heat pump works

A typical heat pump system has three major components:

- heat source collector
- heat pump
- heat distribution system.

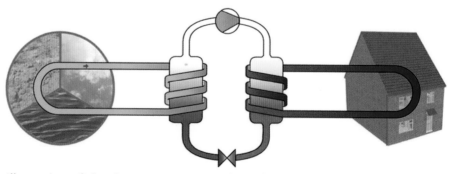

Illustration of the three components of a heat pump

Heat source collector

Heat is mainly provided from three different sources:

- air
- water
- ground.

The Earth stores heat from the Sun around 1m to 2m below the ground surface. In a GSHP, a transfer liquid, which is usually a water and anti-freeze mixture, is pumped through an underground pipe where it absorbs heat from the ground. This heat is absorbed by the transfer liquid and moves to the heat pump. This is known as a closed loop system.

An open loop system draws water up from beneath the ground, extracts the heat from the water and then returns the colder water to the ground. ASHPs extract heat directly from the surrounding air. Water source heat pumps are typically closed loop circuits that extract heat from a body of water.

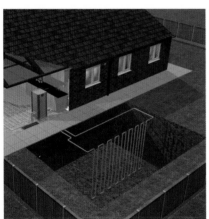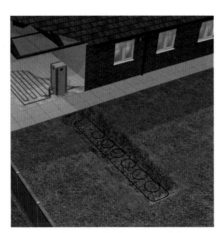

Example of ground source heat pumps

Heat pump

Within the heat pump the **refrigerant** goes through a compression and expansion cycle which increases the heat source temperature.

Stage 1 – Low temperature heat is gained in the **evaporator** where the liquid refrigerant changes to a gas.

Stage 2 – The electrically-driven **compressor** increases the pressure and temperature of the refrigerant and pumps it around the circuit.

Stage 3 – After the compressor, high temperature heat is released in the **condenser** where the refrigerant changes from a gas to a liquid.

---

refrigerant A refrigerant is a compound used in a heat cycle that reversibly undergoes a phase change from a gas to a liquid. Traditionally, fluorocarbons, especially chlorofluorocarbons were used as refrigerants, but they are being phased out because of their ozone depletion effects and more environmentally benign refrigerants have and are being introduced.

evaporator In reference to heat pumps, low temperature heat is gained in the evaporator where the liquid refrigerant is boiled and changes to a gas.

compressor In reference to heat pumps, the electrically-driven compressor increases the pressure and temperature of the refrigerant and pumps it around the circuit.

condenser In reference to heat pumps, high temperature heat produced by the compressor is released in the condenser where the refrigerant changes from a gas to a liquid.

Stage 4 – The fluid in its now liquid state expands to a lower pressure in the **expansion valve**.

expansion valve In reference to heat pumps, refrigerant from the condenser passes through the expansion valve to lower its pressure ready for re-boiling from a liquid to a gas in the evaporator.

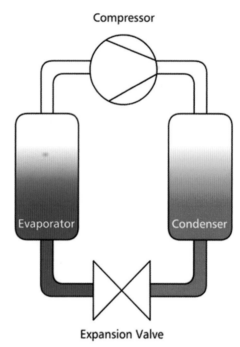

Illustration of the four stages of the compression/expansion cycle

E-LEARNING

Use the e-learning programme to see a demonstration of how a heat pump works.

Heat distribution system

The heat flows via the condenser heat exchanger to an indoor heat distribution system which can be either low temperature radiators or preferably, underfloor heating. If the heat pump needs to produce higher temperatures, for example in a conventional radiator circuit, then its efficiency will be reduced.

MCS MCS is the first product and installer certification scheme to cover all the microgeneration technologies. MCS used to be called the Microgeneration Certification Scheme.

UK AND INTERNATIONAL STANDARDS

The UK MCS accreditation scheme has developed, with its associated trade bodies, the 'Heat Emitter Guide' (HE Guide) which looks at the various radiator and underfloor heating sizes and options for both ASHP and GSHP at different flow temperatures.

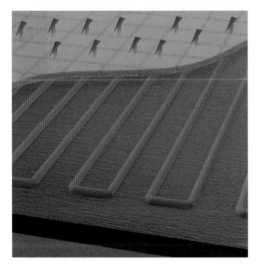

Under floor heating

ACTIVITY 6

Download the pdf of the HE Guide from the MCS website (**www.microgenerationcertification.org**) or search for 'MCS' online and look up the SPF of an ASHP and a GSHP for a flow temperature of 45°C. Looking further at the Guidance table from the HE Guide, for a room heat loss of 80 to 100 W/m2 and the 45°C flow temperature, what are the different heat emitter options that are available and what are the various 'Oversize Factors' and 'Pipe Spacings' that apply?

Advantages vs disadvantages of heat pumps

Air source heat pumps have been very popular in North America and parts of Europe but are now also being widely adopted in the UK, Ireland and many other parts of Europe where the outside air temperature rarely drops below freezing point. They can be fixed to the outside of a building or free standing.

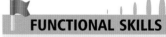

FUNCTIONAL SKILLS

Extract and interpret information from tables

Air source heat pumps are generally cheaper to install than ground source as they do not need the associated ground collector loops. An important consideration with air source heat pumps is that they can be visually intrusive and, depending on location and design, can create some noise. ASHPs are usually a bit less efficient than ground source heat pumps.

Air source heat pump (ASHP)

If well fitted, ground source heat pumps are cheaper to run and more environmentally friendly than air source heat pumps. The ground source loops need to be buried 1m to 2m deep and can be slinky or straight pipe. For a typical domestic installation, a very approximate rule of thumb is that slinkies require 10m to 15m of trench per kW and trenches should be 5m apart.

If space is an issue, the alternative is to use vertical loops or bore-holes but these are more expensive. All ground collectors need careful design, taking into account a wide number of factors. In general, heat pumps are more expensive to install than traditional boiler technology.

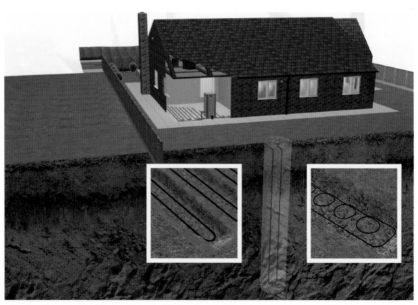

Ground source heat pump (GSHP)

UK AND INTERNATIONAL STANDARDS

Rather than use rules of thumb, heat pumps need to be carefully designed and much more guidance can be found on this process in the UK MCS design methods which is highlighted in the MIS 3005 v3 standard available on the MCS website (**www.microgenerationcertification.org**). The methodology in this document is particularly highlighted as it is the most developed system currently in use. Germany, Austria and Switzerland often use the design methodology in another document called VDI 4640 and its derivatives.

ACTIVITY 7

Of the four pictures displayed, discuss which would probably make a good or poor GSHP installation.

Planning permission for heat pumps

Noise and other standards

UK AND INTERNATIONAL STANDARDS

Ground source or water source heat pumps within the boundaries of a property are classed as permitted development in the General Permitted Development Order which applies in England and Wales. This means that these systems can typically be fitted without any further recourse to planning permission.

General Permitted Development Order 2008 (GPDO) Is an aspect of town and country planning which allows people to undertake minor development under a deemed grant of planning permission, therefore removing the need to submit a planning application. Permitted development is currently set out in the Town and Country Planning (General Permitted Development) Order 1995 (amended in 2008).

Air source heat pumps within the boundaries of a property have more restrictions placed upon them due to the potential noise and visual disturbance they may cause. For example, air source heat pumps are not permitted where they would be louder than 45 decibels 1m from the window of a habitable room in an adjacent property. Full details of the planning permission required is in the General Permitted Development Order which is available online.

Many modern air source heat pumps are fairly quiet in operation. The 45 decibel limit 1m from the neighbour's window is often taken as a general rule. As the rule was introduced in England at the end of 2011, the Government set the level at 42 decibels 1m from the neighbour's window, to be reviewed in a year's time. This noise level limit will vary from nation to nation and from time to time. Check local planning, regulatory and standards requirements in your local area before fitting any new heating or electrical system to make sure your installation adheres to local requirements.

Air source heat pump (ASHP)

ACTIVITY 8

A fairly wide urban terraced house has a small front garden and a large back garden. The house has conventional radiators that are designed to work at a flow temperature of 80°C. What recommendations would you make to the customer so that they could fit a vertical loop or horizontal loop GSHP or an ASHP? Assume the house is in England or Wales (or your own country of origin if you know the local regulations) to offer the best advice.

BIOMASS

An introduction to biomass

The terms 'biomass' and 'biofuels' have come to mean energy derived from a variety of sources including:

- basic wood fuels such as logs, chips and pellets
- anaerobic digestion of waste (breakdown of biodegradable material)
- other vegetable or animal organic matter
- biomass and biofuels can be processed by thermal, chemical or biochemical methods.

For this workbook we are going to concentrate on three different forms of wood fuel as biomass. Biomass is counted as a renewable energy source because the carbon undergoes a natural cycle and the fuel is naturally replaced. There is a significant collection of thermal (combustion), chemical and biochemical processes that can be used to process the many different forms of organic matter that make up biomass and biofuels.

All biomass is inherently **carbon neutral**. Biomass loses its carbon neutral status if fossil fuel is used in its growth, harvesting, processing or transportation. The more fossil fuel that is used, the lower the quantity of renewable energy obtained from the biomass.

> **biodegradable** The chemical breakdown of materials by a physiological environment. The term is often used in relation to ecology, waste management and environmental remediation (bioremediation).

> **carbon neutral** Producing no carbon emissions, or balancing the amount of carbon released with an equivalent amount sequestered or offset.

Wood is an example of biomass

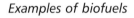

Carbon
Neutral

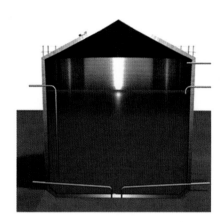

Examples of biofuels

ACTIVITY 9

Coke is a fossil fuel and charcoal is a renewable fuel. Look them up on the Internet and explain the difference between the two substances, what they are made from and name the process used to make charcoal.

Why biomass is carbon neutral

As a biological material, wood contains hydrogen, carbon, nitrogen and oxygen. As part of photosynthesis when the tree was alive, it absorbed carbon dioxide from the air. This same carbon dioxide is released when the tree is burned, causing little or no excess carbon dioxide being released back into the atmosphere, making the process carbon neutral.

Burning trees is a carbon neutral process

E-LEARNING

Use the e-learning programme for more information on why biomass is carbon neutral.

ACTIVITY 10

What are the ingredients required for photosynthesis to occur?

Logs, chips and pellets

Wood fuel is available in a variety of forms and the quality and type of wood will vary widely.

SUSTAINABILITY

If you are purchasing wood fuel as a renewable fuel, it is essential you check that the wood has come from a sustainable source as there is a potentially limited supply of wood.

You will need to compare prices including delivery, the type and size of the wood and also its quantity or volume.

Ensure wood has come from a sustainable source

Logs

Logs have been burned as a fuel for centuries, either as individual logs or in batches. Although open fires look nice, they are very inefficient at producing heat. With an open fire, 80 per cent to 90 per cent of the heat is lost up the chimney and only 10 per cent to 20 per cent actually gets into the room. Using a stove or boiler is much more efficient and would reduce these figures. Additionally, poor combustion of logs in any form will lead to unnecessary pollution.

Logs are delivered in various amounts and need to be stored under-cover but open on the sides. In this way, logs are normally felled in one year and then left to season over the winter in a store. This seasoning of the wood allows the moisture to leave the wood so that when it is subsequently burnt, the energy in the wood is used to heat the building rather than evaporate the residual water content in the wood.

Wood logs

Chips

Trees and other purpose grown crops can be chipped to make a more controllable fuel than logs; these chips are then used in a boiler. As a fuel, the advantages of wood chips are their low cost and their availability in rural areas. Space is less of an issue in rural areas too which means the storage of the chips is less of a problem.

Woodchips require 5m^3 of storage space per tonne depending on the moisture content of the chips. Wood chips are usually stored in a silo or purpose built building close to the boiler. Chips should be a consistent size and shape for ease of handling and processing.

Long, thin chips are a particular nuisance because they tend to block augers and other mechanical devices used to move the chips from the silo into the boiler. Chipped wood will tend to lose its moisture content more rapidly than logs. However, wood chip boilers are flexible in the moisture content of the wood they burn, with moisture contents often around 40 per cent.

Wood chips

Pellets

Wood can be easily made into pellets. First it is ground into sawdust then it is heavily compressed into pellets. The advantage of pellets is that they have a high energy density due to their low moisture content and compact nature and so less storage is required, only 1.5m^3 per tonne.

Disadvantages of wood pellets are the higher cost compared to chips or logs, and the process of making and transporting the pellets can use 20 per cent or more of the pellets' energy.

For all types of wood fuel, fuel quality is very important. The fuel must be the right shape, moisture content and variety of wood for the boiler. For example, if pellets have not been properly compressed, the pellet will crumble and so block the feed system rather than supply useful heat in the combustion chamber.

Another important factor with wood fuel is particulates. Like diesel engines, burning wood can release particles into the local environment. This doesn't typically present an issue in a rural area. However, in an urban environment, it can reduce air quality. Modern wood boilers and stoves are addressing this air quality issue and some are fitted with cyclones and other filters to remove the particles from the flue gases.

HEALTH AND SAFETY

Because burning wood fuel is a combustion technology, all the normal safety implications associated with combustion systems apply and it is vital that an adequate supply of fresh air is provided to the boiler so as to avoid **carbon monoxide** and other safety matters.

carbon monoxide (CO) A colourless, odourless and tasteless gas that is very toxic to humans and animals.

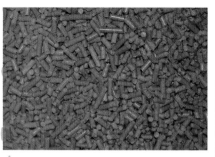

Sawdust pellets

ACTIVITY 11

List the advantages and disadvantages of using logs, chips and pellets.

E-LEARNING

Use the e-learning programme to find out more about logs, chips and pellets.

Log stoves and batch log burners

Logs can be used in stoves and batch burners.

Log stove

Batch log burners

Log stove

Log burning stoves are becoming increasingly popular, particularly because of their attractive appearance, especially in use. At their most basic, they are a metal box with a door for stoking and a flue for removing hot gases.

Modern log burning stoves have secondary and tertiary air combustion systems to burn the volatile gases that are driven from the wood fuel and significantly reduce the unburned carbon.

Efficiency can also be increased if a heat exchanger is added to the flue to remove more of the heat from the hot flue gas. As the fire can only be stoked with relatively large units of fuel i.e. a log, controllability is poor.

There is much discussion about the gains from fitting water heat exchangers to log stoves as these tend to reduce the combustion temperature of the log stove and so can increase sooting within the stove, resulting in more cleaning. Modern lined chimney flues maintain good combustion as they encourage the free flow of air in the flue and so maintain the flow of combustion air in the stove.

Log stove

Batch log burner

One solution to improving the combustion of logs is to use a batch burner. In a batch burner the fire box is filled with logs which are burned quickly and efficiently. The heat is stored in a large tank of water, ready for when it is required by the household so the need for a high level of controllability is avoided.

Batch burners have a fan in the flue which sucks the fire down through the log pile. Some units include systems which monitor the oxygen in the flue gas to ensure optimum operating efficiency by varying the amount of air delivered to the combustion process. Manufacturers claim efficiencies of 80 per cent to 90 per cent for batch log burners.

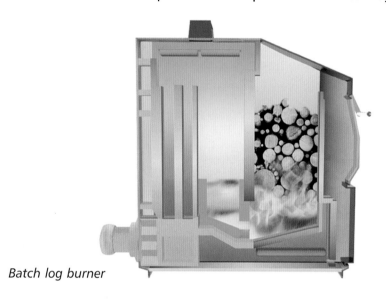

Batch log burner

Woodchip boilers

Woodchip boilers are typically over 40 kW and so are not suitable for domestic use. They are well suited to larger buildings like schools, commercial premises and community applications. As they burn at a higher temperature, they are able to tolerate a higher moisture content than either log or pellet burners.

Woodchip boilers more suited to larger buildings

ACTIVITY 12

The store size for wood is considerably bigger than a bunded oil tank. A ratio of one for oil; three for pellets; and nine for chips is often applied. Using this ratio, if a 1 000 litre oil store is to be replaced by a pellet or a chip store, what size would these stores be? And would an equivalent log store be smaller or bigger than the chip store?

E-LEARNING

Use the e-learning programme to learn more about the different types of burners and boilers.

Pellet stoves and pellet boilers

Pellets can be used in:

- stoves
- boilers.

Pellet stove

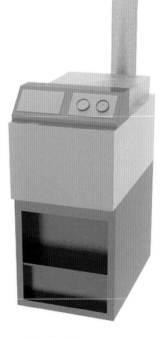

Pellet boiler

Pellet stove

The output of a pellet stove is typically 3 kW to 20 kW. They have a fan which produces warm air for space heating by convection rather than radiating heat. They also have an auger which is a screw that turns to deliver the pellets from the hopper to the burn pot. Both of these require an electrical connection. They have a push button ignition, are 80 per cent to 90 per cent efficient and are very controllable, with some having temperature sensors and timers. They have an integral hopper which is manually fed from bags and an integral ash pan which typically needs emptying once a month or less.

Pellet stove

Pellet boiler

Pellet boilers operate at between 8 kW and 500 kW and are almost completely automatic in their operation. They are divided into three types depending on the direction of the fuel feed:

- top feed
- underfeed
- horizontal feed

The controllability of pellet boilers is excellent and maintenance, cleaning out the ash and checking over the system, is reduced to every five to six weeks.

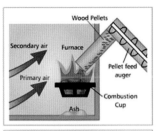
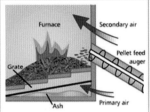
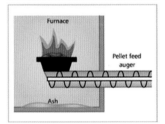

Pellet boiler

Biomass development considerations

UK AND INTERNATIONAL STANDARDS

Biomass systems count as permitted development in the English and Welsh General Permitted Development Order. Exceptions include where a flue exceeds 1m above the roof height excluding the chimney and in listed buildings. In England and Wales, a Building Control Certificate or similar in Scotland will also be required. If in any doubt, contact the local authority before building work starts.

Similar standards will apply throughout Europe. All wood boilers require a flue to make sure the exhaust gases from the combustion process are carried away from the building. These flues need to be a certain height to make sure there is no air blown back down the flue. However, they should not be so tall as to be unsightly. As with all changes to the building, we want to make sure the final solution is attractive to the eye.

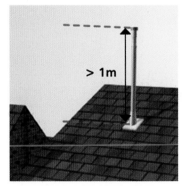

Biomass development consideration needs to follow the UK and International standards

ACTIVITY 13

central heating A system that supplies space and hot water heating to a building from a single heat source through ducts or pipes.

In an urban environment, your existing customer has a gas-fired **central heating** system they are pleased with and don't want to change. However, they would like to add a renewable energy heating option to their property. They contact you to ask about their existing chimney and what options they have to fit a wood-fired additional heat source that will be more environmentally friendly than their existing open fire. What options might you recommend?

MICRO COMBINED HEAT AND POWER

An overview of micro CHP

Micro Combined Heat and Power (mCHP) Micro-CHP is an extension of the now well established idea of cogeneration (the use of a heat engine or a power station to simultaneously generate both electricity and useful heat). It is one of the most common forms of energy recycling to the single/multi family home or small office building.

Micro Combined Heat and Power or Micro CHP is the simultaneous production of heat and electricity in individual homes. The Micro CHP unit replaces the existing central heating boiler and provides heat and hot water as usual and also provides some of the home's electricity needs as well.

Any excess electricity can be exported to the national grid. Although Micro CHP units produce relatively small amounts of electricity, the potential lies in the large number of systems which may eventually be installed in the millions of homes in the UK and other European countries.

Optimistic predictions hope that eventually, Micro CHP may provide around 20 per cent of the UK's electricity generating capacity. Micro CHP is often written as **mCHP**.

House cross section, showing micro combined heat and power unit.

How does micro CHP work?

Natural gas is consumed in a **Stirling engine** or other Micro CHP unit to provide heat and electricity for use within the home. A total of around 70 per cent to 80 per cent of the energy value of the gas is converted into heat, mainly in the form of hot water which is used for space heating and domestic hot water as in a normal central heating system. Between 10 per cent and 25 per cent is converted into electricity, and the remaining 10 per cent to 15 per cent is lost in the flue gases.

Although the efficiency of a Micro CHP system is similar to a boiler system, the electricity produced has a much higher value than heat. It is the value of this electricity which covers the investment cost of the micro CHP unit and provides an overall saving. Before considering a Micro CHP system, you should undertake an economic viability assessment taking into account incentive payments from the **Feed in Tariff** (FiT).

In Britain, the national Government pays householders for producing electricity from mCHP units. Check with local conditions to see if your local, regional or national Government makes FiT payments on mCHP.

As well as Stirling engines, micro CHP and larger CHP units use other engines such as internal combustion engines and fuel cells. All combined heat and power systems require a fuel and an engine that produces both heat and power to make electricity.

Stirling engine A Stirling engine is a heat engine that operates by cyclic compression and expansion of air or other gas, the *working fluid*, at different temperature levels such that there is a net conversion of heat energy to mechanical work.

Feed-in-Tariff (FiT) FIT or renewable energy payments is a policy mechanism designed to encourage the adoption of renewable energy sources and to help accelerate the move toward grid parity. Energy suppliers will make two types of payments to householders and communities. One payment will be to those who generate their own electricity (Generation Tariff) from renewable or low carbon sources. The other will be for the excessive electricity generated which has been exported back to the grid (Export Tariff).

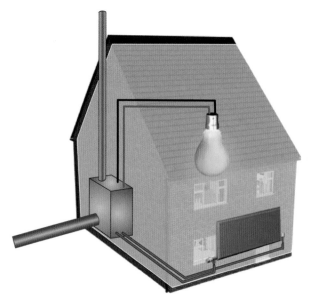

Illustration of the working of a micro CHP

ACTIVITY 14

Use the Internet to find out who invented the Stirling engine and when? If a petrol or diesel engine is an internal combustion engine, what is a Stirling engine classed as? A fuel cell is a device that converts the chemical energy from a fuel into electricity. Can it be powered by hydrogen, natural gas and methanol? Can all these engines drive CHP and mCHP systems?

Planning permission for micro CHP

UK AND INTERNATIONAL STANDARDS

Planning permission is not normally required for the internal work when installing a Micro CHP system in a house. The external flue will normally be classed as permitted development in the General Permitted Development Order with a few exceptions. Listed buildings and buildings in a conservation area or World Heritage site will have further restrictions and if in doubt, you should always check with your local planning authority before a flue is fitted.

Building regulations also apply to other aspects of the work such as electrical installation and plumbing work. These rules will probably apply on a fairly universal basis across many countries as all combustion systems require a flue to remove the exhaust gases from the property and building standards will all require high quality and safe workmanship. As ever, always check local planning and building requirements before engaging in any building services work.

An unusual aspect of mCHP is that with this building services system, the property can be feeding power back into the grid as well as accepting power from the grid. Most countries are now, as a result of PV becoming commonplace, getting used to this process.

ACTIVITY 15

What building services skills will be required to fully connect a mCHP system? Who will have to be informed of the installation of a mCHP unit as well as taking into consideration any planning and building control/standards requirements?

CHECK YOUR KNOWLEDGE

1. Which of the following is correct?

☐ a. Northern Europe has minimal solar energy.

☐ b. Solar power can only be used for hot water and swimming pool heating.

☐ c. Evacuated tubes are much better than flat plates.

☐ d. Twin-coil and two cylinder layouts both work well with solar.

2. Which of the following is correct?

☐ a. When fitting a solar system, building and water regulations should be followed.

☐ b. A competent plumber can fit the whole solar system.

☐ c. Solar only works with vented or unvented cylinders.

☐ d. Active systems are pumped and passive systems use pressure.

3. Which of the following is correct?

☐ a. All solar panel installations which protrude less than 200mm above the roofline automatically qualify for planning permission.

☐ b. Building control is the process of meeting local appearance requirements.

☐ c. If panels are fitted east- or west-facing, the panel area should be increased by around 25 per cent.

☐ d. If the pipework between solar collector and cylinder is short, the pipes don't need to be insulated.

4. Which of the following is correct?

☐ a. Heat pumps are not classed as renewable energy.

☐ b. Heat pumps only save energy if the system is well installed, commissioned and maintained.

☐ c. A Heat pump has two major components, a heat source and a heat distribution system.

☐ d. Within the heat pump the refrigerant goes through a compression and expansion cycle which increases the heat source temperature.

5. Which of the following is incorrect?

☐ a. Biomass and biofuels can be processed by thermal, chemical or biochemical methods.

☐ b. All biomass is inherently carbon neutral. Biomass loses its carbon neutral status if fossil fuel is used in its growth, harvesting, processing or transportation.

☐ c. With an open fire, 50 per cent to 60 per cent of the heat is lost up the chimney.

☐ d. Long, thin chips are a particular nuisance because they tend to block augers.

☐ e. For all types of wood fuel, fuel quality is very important. The fuel must be the right shape, moisture content and variety of wood for the boiler.

6. Which of the following is incorrect?

☐ a. Micro CHP is the simultaneous production of heat and electricity in individual homes.

☐ b. Optimistic predictions hope that eventually, Micro CHP may provide around 20 per cent of the UK's electricity generating capacity.

☐ c. Natural gas is the only fuel that can be used in a micro CHP unit. Petrol, diesel or another fuel is not suitable.

☐ d. Micro CHP and larger CHP units use engines such as Stirling engines, internal combustion engines and fuel cells to make heat and power.

7. **Which of the following statements are advantages and disadvantages of solar thermal water heating? Fill in the table stating Advantage or Disadvantage.**

	Advantage or Disadvantage
Expensive with a high from end cost	
Fits virtually every building	
Should be near the distribution system	
Only useful for short term storage	
Very flexible	
Zero carbon technology	

8. **Why is biomass fuel carbon neutral?**

☐ a. Because wood does not contain carbon.

☐ b. Because when wood burns it only releases as much carbon dioxide as the tree absorbed in its lifetime.

☐ c. Because there is an unlimited supply of biomass fuel.

9. **What does Micro CHP stand for?**

☐ a. Micro Continuous Heat and Power.

☐ b. Micro Circuit Heat and Power.

☐ c. Micro Combined Heat and Power.

Chapter 3

ELECTRICITY PRODUCING TECHNOLOGIES

LEARNING OBJECTIVES

By the end of this chapter you will be able to:

- describe the different forms of electricity producing technologies

- explain where solar photovoltaic panels would be applicable

- explain where wind energy would be applicable

- explain where hydroelectricity would be applicable

- list the advantages and disadvantages of each type of electricity producing technology

- list the building location, features, fabric, regulations and planning permission that would apply to each technology

PHOTOVOLTAIC ENERGY

An overview of photovoltaic cells

The word photovoltaic comes from photo which means light and voltaic which means electricity. Photovoltaic or PV cells convert the energy from the Sun directly into electricity. A group of cells connected together is called a module and a group of modules packaged into a frame is more commonly known as a solar or PV panel. A group of PV panels is known as an array.

In the UK and Ireland we have a useful source of solar energy with around half of this energy being diffuse or indirect sunlight. Electricity can still be produced on cloudy or overcast days but the highest output will always come from the strongest, unobscured sunlight.

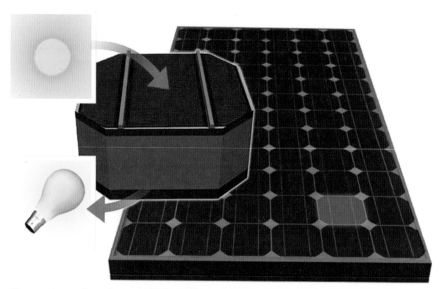

Illustration of a photovoltaic cell

PV cells have many uses. They can be used in situations where there is no access to, or it is impractical to use, mains electricity, and to power appliances from satellites to watches. They are frequently used in the developing world to power water pumps and more recently, due to government support, the UK domestic market is rapidly growing.

Photovoltaic cells

Types of photovoltaic panels

There are three different types of PV panel:

- monocrystalline
- poly or multicrystalline
- amorphous or thin film.

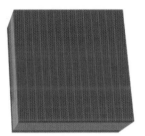

Monocrystalline *Poly or mutlicrystalline* *Amorphous or thin film*

Monocrystalline

Monocrystalline panels are the most expensive. The silicon is cut from cylinder shaped ingots and so does not completely cover a square PV cell module without wasting silicon. This leaves uncovered gaps at the four corners of the cells. They are the most efficient, however, at 15 per cent to 18 per cent efficiency.

Monocrystalline

Polycrystalline

In poly or multicrystalline panels, the silicon is either cut from cast square ingots or drawn in flat thin ribbons from molten silicon. They are less expensive to produce than monocrystalline panels but also less efficient at 14 per cent to 15 per cent efficiency.

Polycrystalline

monocrystalline In Solar PV panel terms, monocrystalline panels are the most expensive. A silicon ingot is extracted from a liquid vat of silicon and cools into a single crystal cylinder. The cylinder shaped ingots are then cut into 0.3mm thick slivers and arranged into PV panels. They are the most efficient, however, at 15 per cent to 18 per cent efficiency.

amorphous In Solar PV panel terms, amorphous or thin film coatings do not have a regular crystalline structure and as the material is coated onto a surface, it would appear to have a haphazard structure under a microscope. Less light absorbing material is required to create a PV cell, hence these amorphous panels are often also called thin film. This often makes them the cheapest to produce but also leads to reduced efficiency of around 5 to 7 per cent. They have also become popular due to their flexibility, lighter weights, and ease of integration. Some more modern thin film panels come in layers which increase both efficiency and cost.

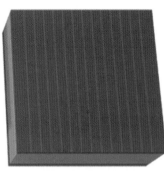

Amorphous/thin film panels

Amorphous

Amorphous or thin film panels reduce the amount of light-absorbing material required to create a PV cell. This makes them the cheapest to produce but also leads to reduced efficiency of around 5 per cent to 7 per cent. They have also become popular due to their flexibility, lighter weights and ease of integration. Some more modern thin film panels come in layers which increase both efficiency and cost.

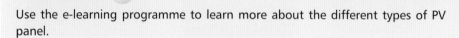

E-LEARNING

Use the e-learning programme to learn more about the different types of PV panel.

ACTIVITY 16

The significant majority of currently manufactured PV panels are made from silicon. Use the Internet to investigate and list two other materials that PV panels are made from.

Stand alone vs grid connected systems

There are two different types of photovoltaic system:

- stand alone
- grid connection.

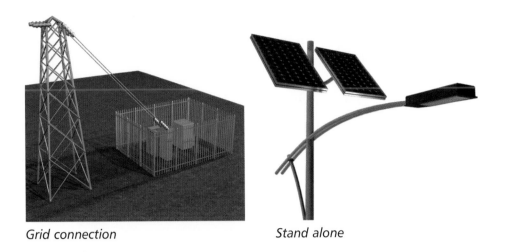

Grid connection   *Stand alone*

Stand alone system

Stand alone systems are used where there is no access to mains electricity or it is impractical to use it. In developing countries, the electricity grid tends to be confined to the main urban areas and does not reach rural areas. In this case it can be the best and least expensive way of providing electricity to these areas.

In the developed world it is now not unusual to see PV powered street signs, phone kiosks and weather monitors. You can even buy stand alone PV chargers for mobile phones and MP3 players.

Examples of stand alone systems

Grid connection

A mains connection means that the grid can provide additional electricity when there is insufficient sunlight to meet the electricity demand of the user. It also works in reverse in that excess electricity can be passed back via the grid to power other local houses and so power stations burn less fossil fuel.

In the UK, grid connected PV and other renewable electricity gains both a **generation tariff** for all the electricity produced and an **export tariff** for all the electricity that is sold to the grid. This is called a Feed-in-Tariff (FiT). There are about 22 different FiT schemes in operation around the globe and each scheme has slightly different payment rates and metering methods so it's important to check your local area for its needs.

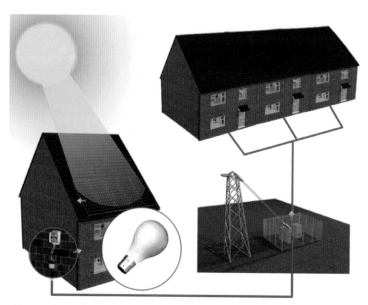

Illustration showing how a grid connection system works

Inverters and battery charge controllers

PV panels produce DC power which is similar in nature to the DC power from a battery. For stand alone PV systems connected to a battery, the system will require a Battery Charge Controller (BCC) to manage the electrical interface between the battery and the PV panel. This BCC disconnects the PV panel from the battery when the battery is fully charged and disconnects the electrical load from the battery when the battery is fully discharged, so protecting and extending battery life.

The mains electricity grid is run on AC power and so PV panels cannot be directly connected to this grid. Therefore, an inverter is used in a mains-connected PV circuit to change the DC power to AC power. This inverter has several safety features.

E-LEARNING

These safety features and much more detail on PV systems are discussed in the Solar PV e-learning programme.

ACTIVITY 17

A grid disconnected caravan has been equipped with a PV array, a battery charge controller, a solar compatible battery and a series of DC-powered lights and a DC-powered radio. However, it also has a 240V AC-powered TV. What will this system also need if the occupants want to watch TV this evening?

Planning permission and building control/standards for photovoltaic panels

UK AND INTERNATIONAL STANDARDS

In England and Wales, photovoltaic panels are counted as permitted development under the General Permitted Development Order and do not usually require planning permission. Exceptions apply if the panels extend more than 200mm over the existing roof plane, if the building is listed or if it is in a conservation area or World Heritage site. Similar planning standards are likely to apply in various different regions as Governments tend to be very keen to make sure as much PV as possible can be fitted in their locality.

Again, within England and Wales, a building control certificate is required for the electrical circuit. You may also need building control approval to make changes to the roof or mounting location of the panels. Both competent roofers and electricians are required to fit PV systems and good installation practices must be followed. Therefore always consult your local authority before beginning installation of photovoltaic panels to make sure you have complied with all the local building requirements.

Photovoltaic panels are permitted development under the General Permitted Development Order

WIND ENERGY

An overview of wind energy

Wind turbines are categorized as small, medium and large, and harness the power of the wind to generate electricity.

Systems of up to 50 kW are known as small wind turbines. They are suitable for generating electricity to power the lights and electrical appliances in a typical home, school or small business. Forty per cent of all the wind energy in Europe blows over the UK, which makes it an ideal country for small wind turbines.

Other North European countries such as Denmark and Northern Germany also have good wind speed regimes as do many parts of Spain and Portugal. Wind tends to be far more site specific than solar renewable technologies as it has more variability, so it's important to evaluate the site carefully to ensure the predicted returns are realistic.

Examples of wind tubines

Wind turbines less than 50 kW are sometimes called micro, mini or small wind turbines. The exact definitions vary from company to company and country to country. For economies of scale, within large wind turbines, there is a trend towards machines of several **megawatts** in size. However, this is a process of diminishing returns and the biggest on the market is around 7.5 MW whilst a typical large wind turbine is up to 2.5 MW.

> **megawatts (MW)** The megawatt is equal to one million watts. Many events or machines produce or sustain the conversion of power on this scale.

How do wind turbines work?

Wind turbines use blades to catch the wind. When the wind blows the blades are forced round, driving a turbine which generates electricity. The stronger the wind, the more electricity is produced. There are two types of small wind turbine: Mast-mounted which are free standing in a suitably exposed position, typically around 2.5 kW to 6 kW, and roof-mounted which are smaller and can be installed on the ridge of a house roof typically around 1 kW to 2 kW in size.

If the small wind system is connected to the national grid then excess electricity can be passed back to the grid, providing a cost saving. If the system isn't connected to the national grid then excess electricity can be stored in batteries to use when there is no wind.

E-LEARNING

See the e-learning programme for a demonstration of a wind turbine.

Excess electricity produced can be stored in batteries to use when there's no wind

ACTIVITY 18

Modern wind turbines are typically used to generate electricity. However, wind turbines have been used for many centuries for their mechanical power. Suggest two traditional mechanical uses of wind turbine power.

Small wind turbines are more suitable for rural areas

Is a small wind solution suitable?

There are a number of things that will determine whether a small wind turbine is suitable for a particular situation. Small wind turbines work best in exposed locations without turbulence caused by large obstacles like buildings, trees or hills. To be effective they need an average wind speed of no less than 5m per second and so are particularly suited to rural locations where mains electricity is sometimes unavailable.

UK AND INTERNATIONAL STANDARDS

Small wind systems might require planning permission, so the opinion of the local authority should be sought before installation begins, particularly if noise may be an issue.

A final consideration may be whether to install a vertical or horizontal axis wind turbine. Vertical axis wind turbines are not as efficient as horizontal axis wind turbines but do perform better in low wind situations as they can take the wind's energy from any direction.

ACTIVITY 19

The power available on a wind turbine is the cube (rather than the square) of the wind speed. Therefore, how much more energy does the wind have at 10 m/s as compared to 5 m/s?

MICRO HYDROELECTRICITY

Introduction to micro hydroelectricity

Hydro power uses the movement of water to turn a turbine, which generates electricity. The faster the water flows and the more water there is, the more electricity can be generated. The energy extracted from the water is the difference in height between the source and the water's outflow and the volume of water that will flow. This height difference is called the **head**.

Micro **hydroelectricity** describes an installation with a fall of water less than 5m, for example river or stream flows, to produce energy. These applications do not need to dam water to create head so can create electricity with a minimal impact on the environment.

head In Micro Hydroelectricity terms, head is the term used to describe the difference in height between the water source and the water outflow.

hydroelectricity Electricity generated by hydropower, i.e., the production of electrical power through the use of the gravitational force of falling or flowing water. It is currently the most widely used form of renewable energy.

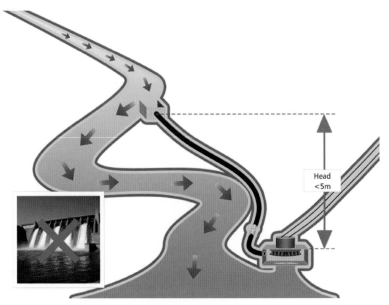

Illustration of micro hydroelectricity

Constructing a micro hydroelectricity system

Very few locations are suitable for a micro hydroelectricity system. They need a river or stream that flows year round and each one is custom made for its location, making it expensive. Where they are suitable however, they provide a quiet, steady and reliable source of power.

The common elements of all hydroelectric systems are the same:

- a supply of water is needed
- an intake structure to screen out floating debris and fish
- a pipe or **penstock** to move the water to the turbine
- a controlling valve to regulate the flow and the speed of the turbine

The turbine then converts the flow and pressure of the water to energy and the water emerging from the turbine returns to the natural watercourse along a **tailrace** channel.

penstock A penstock is a sluice or gate or intake structure that controls water flow, or an enclosed pipe that delivers water to hydraulic turbines and sewerage systems.

tailrace (Micro Hydroelectricity) In Micro Hydroelectricity terms, a tailrace is the channel of water flowing back to its natural watercourse.

UK AND INTERNATIONAL STANDARDS

An environmental assessment is essential if you are intending to install a micro hydroelectricity system, so it is recommended that you consult the planning authorities, the Environment Agency and Natural England before installation begins. This same advice will apply whatever country you are in.

There will always be interested parties such as environmental bodies and local authorities that will need to be consulted before a hydro system can be installed. However, if a reasonable body and flow of water is available, it is definitely worth making the effort to follow through on a micro hydro installation.

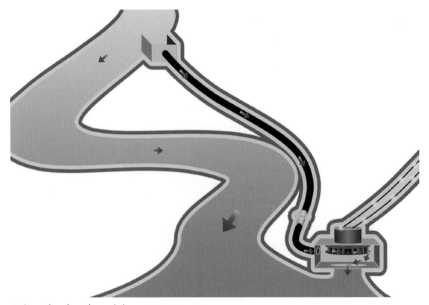

Micro hydroelectricity system

E-LEARNING

See the e-learning programme for a demonstration of micro hydroelectricity.

ACTIVITY 20

Traditional water wheels tended to use flat blades or buckets to generate their mechanical power. Most modern machines feed the water down a channel or a tube to a turbine to make the mechanical power. Please name the technical term used in hydro technology for the tube that feeds the water into the turbine (clue: use the Internet to look up hydroelectricity and you should find it named on some diagrams) and also state the benefit of a modern turbine as compared to a traditional water wheel.

CHECK YOUR KNOWLEDGE

1. Which of the following is correct?

☐ a. A group of PV panels is known as a module.

☐ b. PV panels have in-built storage and so power appliances day and night.

☐ c. PV comes in mono, multi and amorphous crystalline layouts.

☐ d. Monocrystalline are the most efficient.

2. Which of the following is correct?

☐ a. There are two basic layouts of PV, grid connected and stand alone.

☐ b. All FiTs have a generation tariff and an export tariff.

☐ c. The inverters only function is to connect DC to AC.

☐ d. The BCC disconnects the PV panel from the battery when the battery is fully discharged and disconnects the electrical load from the battery when the battery is fully charged.

3. Which of the following is correct?

☐ a. All wind turbines below 60 kW are classed as small.

☐ b. All of Northern Germany is good for producing wind power.

☐ c. Wind can be grid connected or stored in batteries.

☐ d. To be effective, a wind turbine needs an average wind speed of no less than 6 m/s.

4. Which of the following is correct?

☐ a. Small wind turbines work best in exposed locations with turbulence caused by large obstacles like buildings, trees or hills.

☐ b. Wind turbines require planning permission only for appearance issues.

☐ c. Vertical axis wind turbines are as efficient as horizontal axis wind turbines.

☐ d. Vertical axis wind turbines can take the wind's energy from any direction.

5. Which of the following is incorrect?

☐ a. Many locations are suitable for a micro hydroelectricity system.

☐ b. A year round supply of water is required for hydropower.

☐ c. All hydro systems have a pipe or penstock to move the water to the turbine.

☐ d. Hydro systems use a control valve to regulate the water flow and speed.

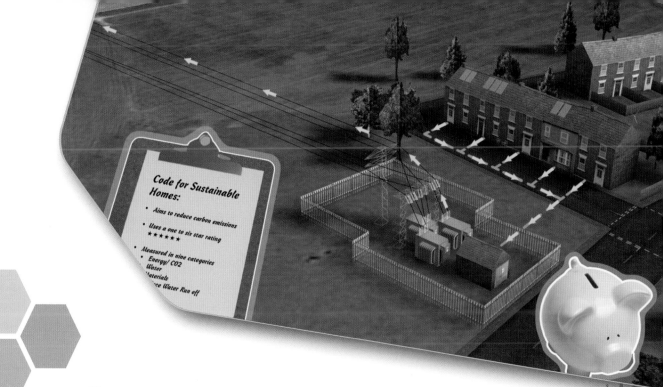

Chapter 4

INCENTIVES AND POLICY

LEARNING OBJECTIVES

By the end of this chapter you will be able to:

- list the types of financial support and incentives available for renewable energies

- list the nine categories in the Code for Sustainable Homes

- list the main accreditation schemes for renewable energy products and installers

- discuss how this typical policy framework might be used in another country

INTRODUCTION

In this chapter, we will discuss what constitutes a renewable energy policy framework that will facilitate the movement from a fossilfuel-powered energy framework to a renewable energy-powered policy framework. We will in particular be looking at the policy framework used in England and Wales, two countries that make up part of the UK.

Many countries have a federal or similar structure. Due to this, it is common to find a whole host of different renewable energy policy frameworks in different regions within the same country. However, the trend that is common to most countries is the move from fossil fuels to renewable energy.

UK AND INTERNATIONAL STANDARDS

The main driver across the EU for renewable energy is the RES directive which calls for Europe to source 20 per cent of its energy from renewable sources by 2020. The diagram shown below highlights the change required in each country within the EU and why this course is focusing on the English and Welsh policy framework, as the UK needs to make the largest percentage improvement over the next years.

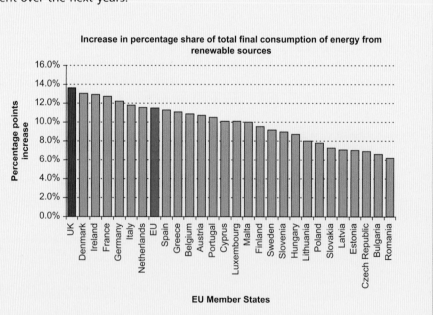

AN EFFECTIVE POLICY FRAMEWORK

When we study energy policy, there are three areas that the policy will look to influence:

- electricity
- heat (which includes cooling)
- travel.

Heat is the biggest consumer of energy, followed by electricity and then travel. About 49 per cent of Europe's energy is consumed providing heating and cooling. A good policy framework needs to address all three areas if the change is comprehensive.

A complete policy framework will include the following three elements:

- Financial incentives in the form of grants, obligations or direct incentives to move existing buildings and consumers to renewables
- Regulation to legislate that new buildings and potential consumers of energy have to use renewables rather than fossil fuels
- Accreditation schemes to make sure that quality is maintained during the change process

These three policy elements must be supported by the 'flanking measures' of good training schemes and an effective public and business awareness campaign.

ACTIVITY 21

If heat is the biggest consumer of energy, why is electricity the biggest emitter of carbon dioxide? To provide some further useful information on this process, in the UK:

- natural gas (the main fuel used for heating) has a carbon dioxide rate of 0.205 $kgCO_2$/kWh
- electricity has a carbon dioxide rate of 0.541 $kgCO_2$/kWh.

Continued on next page

Assume that 49 per cent of consumption is heat, 31 per cent of consumption is electricity and the last 20 per cent of consumption is transport.

Use simplifications to obtain a quick answer. Please note that the 0.541 $kgCO_2$/kWh for electricity is a UK rate. Countries such as Germany with lots of coal-fired power stations tend to have a higher rate and countries such as Sweden with lots of hydro power tend to have a lower rate.

Renewable Heat Incentive (RHI) A scheme which provides financial support to those who install renewable heating technologies such as ASHP or GSHP, solar thermal and biomass boilers, etc. The scheme supports heating at all scales including domestic, commercial, industrial and public sectors.

FINANCIAL FUNDING

In England, Scotland and Wales, there is a **Renewable Heat Incentive (RHI)** and a Feed-in-Tariff (FiT), which incentivize the uptake of renewable energy for existing buildings, as well as consumers of heat or electricity respectively.

Renewable heat incentive

The Renewable Heat Incentive provides financial support for households, businesses and other users that install renewable heating technologies qualifying for support under the scheme. The key aspects of the Renewable Heat Incentive are that the scheme supports a range of technologies and supports heating at all levels up to 5 MW, from households to industrial processes in large factories. Incentive levels are calculated to bridge the financial gap between the cost of conventional and the cost of renewable heat systems. Full information can be found at the Department of Energy and Climate Change website.

FUNCTIONAL SKILLS

Calculations in a practical context

ACTIVITY 22

One of the observations about Government policy is that it is subject to change, especially in a democracy when Governments change. In the above paragraph, when the policy was under development, it was believed that the maximum size of RHI installation was 5 MW. In reality, now that the policy is in place, there is no upper limit on the size of installation. We have left this difference in place to stress the importance of checking the detail of the local policy framework before engaging in a new renewable installation so that you or your customer fully benefits from the local policies.

To make an incentive scheme work, the amount of heat a consumer uses has to be either measured or estimated so that the end user can receive a payment. However, if the heat is measured and the incentive payment is greater than the cost of the fuel, what is to stop the end user from wasting heating by say, opening the windows?

Feed-in-tariffs

These British Feed-in-Tariffs or FiTs are designed as an incentive for small scale to medium scale, low and **zero carbon** electricity generation. They will allow many people to invest in technology in return for a guaranteed payment for the electricity they generate and this scheme works alongside the Renewable Heat Incentive. Technologies eligible include **solar PV**, wind, geothermal, hydro, wave, tidal, biomass and small scale micro combined heat and power. More information can be found at the Department of Energy and Climate Change website.

zero carbon Producing no carbon emissions.

solar PV A Solar Photovoltaic System uses solar cells to convert light into electricity.

FUNCTIONAL SKILLS

Calculations in a practical context

There are a collection of over 22 different FiT policies in use around the world, all with similarities and differences. The sheer number of these schemes demonstrates how effective they have become in incentivizing the growth of renewable electricity technologies.

Illustration of how Feed-in Tariff (FiT) works

ACTIVITY 23

In the UK, the FiT is paid out mainly for generation of power, and the generation meter is inserted as per the diagram shown. In Germany and many other countries, the FiT is paid out for the amount of electricity exported. Where would the export meter be fitted on the diagram shown?

REGULATIONS

SUSTAINABILITY

Regulations are an effective way to legislate so that for all new construction and engineering projects, the power should come from renewable rather than fossil fuel sources. In many parts of Europe, the Passivhaus scheme has been taken forward as a design standard to encourage very low carbon buildings. In England and Wales, the Government addressed this issue with an environmental policy called the Code for Sustainable homes, which looked to address not just the energy issues, but also other environmental issues addressing the occupants' lifestyle.

Code for sustainable homes

The Code for Sustainable Homes is the English and Welsh national standard for the sustainable design and construction of new homes. It aims to reduce our carbon emissions and create homes that are more sustainable. The Code measures the sustainability of a new home against nine categories of sustainable design. It uses a one to six star rating system to rate the whole home as a complete package.

The nine categories are:

- energy or CO_2 water
- materials
- surface water run-off
- waste
- pollution
- health
- well being
- management
- ecology

As the Code for Sustainable Homes only addresses energy matters, it doesn't directly drive the adoption of renewable energy. Many of the Code's requirements can be met with energy efficiency measures and this is a good thing, as energy efficiency matters should be addressed before renewable energy is installed.

In practice, we want both energy efficiency and renewable energy. Therefore, certain local authorities across the UK, with the first being the London Borough of Merton, have adopted a renewable energy charter than requires all new construction over a certain size to have a percentage of its energy supplied by renewables. Similar policies have been used across many different countries.

A house with 5 out of 6 stars from the star rating system

The Merton Rule

UK AND INTERNATIONAL STANDARDS

The Merton Rule is named after Merton Council in Surrey who were first to adopt the planning policy that required new developments to generate at least 10 per cent of their energy needs from on-site renewable energy equipment. The most commonly accepted threshold for the rule is 10 homes or $1000m^2$ of non-residential development.

It is important to note that the Merton Rule covers all buildings and not just homes. As the huge number of planned houses gets built they will be accompanied by new services including schools, supermarkets, shopping centres, office blocks and leisure centres. It is essential that these heavy energy users also play their part in contributing to the Government's renewable energy and climate change strategies and targets.

In 2008, the UK Government published its central planning guidance that required all planning authorities in England to adopt a Merton Rule Policy. Regional and devolved nation differences apply, so the latest information should be consulted on Government websites.

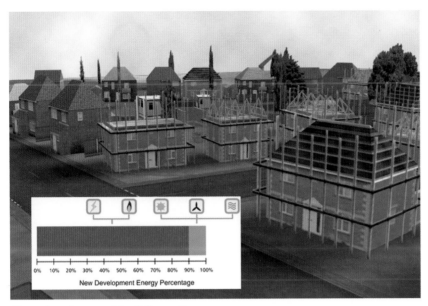

New development building site

ACCREDITATION SCHEMES

UK AND INTERNATIONAL STANDARDS

Good accreditation schemes support the growth of renewable technology. The most widely used and known accreditation scheme across Europe is the **Solar Keymark**. This scheme ensures the quality of solar thermal collector design and production and is accepted across all European nations including the UK, where it is accepted as an equivalent to the national Microgeneration Certification Scheme or as it is more commonly known, MCS. Here below we will discuss some of the particular features of this British scheme.

Solar Keymark The Solar Keymark is an official quality mark of the European Committee for Standardization (CEN), developed with ESTIF to overcome the varying testing and certification requirements in different EU countries. It accredits the quality of solar thermal collectors and systems.

Microgeneration certification scheme

MCS is an independent scheme that certifies microgeneration products and installers to consistent standards. It is designed to evaluate microgeneration products and installers, providing greater protection for consumers and has support from the Department of Energy and Climate Change. When used in conjunction with a **Competent Persons Scheme** or **CPS**, installers are able to qualify for grants and notify installations according to building regulations. Solar Keymark is a voluntary certification scheme supported by the European Solar Thermal Industry Federation to help users to choose quality solar collectors and systems. It is an equivalent scheme to the MCS mark.

Competent Persons Scheme (CPS) A Competent Person is a business which has been registered with an approved scheme by the Department for Communities and Local Government (DCLG). A Competent Person is adjudged to be sufficiently competent to self-certify that all work complies with the relevant part of the Building Regulations (England and Wales).

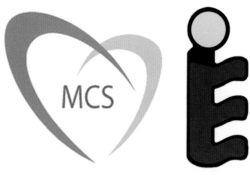

Microgeneration Certification Scheme

ACTIVITY 24

Qualicert is a pan European scheme which aims to unify installer accreditation across various European member states. Please list four Qualicert project partners and also list any (maximum of four) heating and/or electrical accreditation schemes that might also apply in your own country (clue: think of schemes for solid fuel, oil or gas).

1.

2.

3.

4.

CHECK YOUR KNOWLEDGE

1. Which of the following list should not be a part of Government policy; Renewable energy policy will look to influence:

☐ a. Electricity

☐ b. Fuel

☐ c. Heat (which includes cooling)

☐ d. Travel

2. **A complete policy framework will include 3 elements supported by two flanking measures. Which of the following is not part of these elements or measures?**

☐ a. Financial Incentives

☐ b. Regulation

☐ c. Accreditation Schemes

☐ d. Grants

☐ e. Public Awareness campaigns

☐ f. Training

3. **Which of the following is correct?**

☐ a. FiT stands for Feed in Temperature.

☐ b. RHI stands for Renewable Heat Interest (rate).

☐ c. The Code for Sustainable Homes is an energy regulation.

☐ d. The Merton Rule required new developments to generate at least 10 per cent of their energy needs from on-site renewable energy equipment.

4. **Which of the following is correct?**

☐ a. The most widely used and known accreditation scheme across Europe is the Solar Keymark.This scheme ensures the quality of the solar installation.

☐ b. In the solar thermal sector, the UK national equivalent to Solar Keymark is MSC. MSC stands for Microgeneration Scheme for Certification.

☐ c. The UK national accreditation scheme is designed to evaluate microgeneration products and installers, providing greater financial returns for consumers.

☐ d. Solar Keymark is a voluntary certification scheme supported by the European Solar Thermal Industry Federation to help users to choose quality solar collectors and systems.

5. **What is the biggest system funded by the renewable heat incentive?**

☐ a. 5 MW

☐ b. 7 MW

☐ c. 10 MW

6. **What can MCS be used in conjunction with, to allow installers to notify installations according to building regulations?**

☐ a. Certified Persons Scheme

☐ b. Certified Peoples Scheme

☐ c. Completed Persons Scheme

Chapter 5

SELECTION PROCESS

LEARNING OBJECTIVES

By the end of this chapter you will be able to:

- list the items to be checked in an external evaluation and an internal audit of a building

- list the items to be checked in a customer interview

- select appropriate renewable energy solutions for a building

EVALUATION AND AUDIT

Introduction to the selection process

Before considering any form of renewable energy solution in a building there is a three stage selection process that must be completed. This starts with an external evaluation of the building, is followed up by an internal audit of the building and finally concludes with a customer interview.

External evaluation of the building

External site evaluation

The first part of the selection process is to conduct an external site evaluation of the building.

Conducting an external site evaluation

Points to consider in an external site evaluation

The building type

What is the age of the building?
What is the construction of the building?
What is the total floor area of the building?

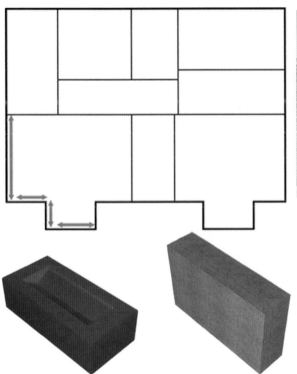

Factors to consider: age, construction and total floor area

The building orientation

What is the pitch and direction of the roof?
Is there any roof shading?
Are there any properties nearby that could be affected by noise?
What speed and direction is the prevailing wind?

The orientation of the building

Outbuildings

Are there any outbuildings?
What is the total size of the outbuildings?
What is the construction of the outbuildings?

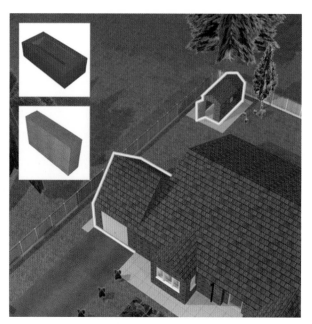

Outbuildings such as garages; sheds; barns etc.

The construction of the building

What are the walls made of?
What is the roof made of?
Are there any chimneys or flues?
What type of windows, doors and glazing does the building have?

Building's construction: walls, roof, chimneys, windows, doors etc.

The land

What is the total size of the land belonging to the building?

What is the terrain of the land like?

Is there sufficient access for machinery to complete the installation?

What is the ground type?

Are there any rivers or streams nearby?

Land considerations

ACTIVITY 25

During an external site evaluation, can you name the two typical wall constructions used in the UK (and across the rest of Europe) and another type of wall construction that is quite commonly used in several countries in Europe. Likewise, can you name three typical types of slate or tile roof?

Internal audit

The external site evaluation is followed by an internal audit of the building.

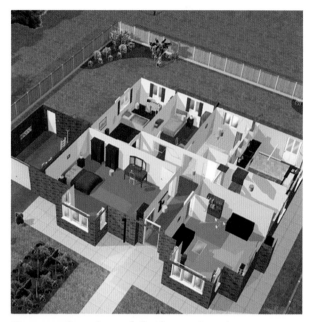

Cross section view of the house interior

Elements to consider during an internal site audit

Metering

Does the building have gas and/or electricity meters?
Where are the meters located?
What type of meters does the building have?
Is there access for export metering?

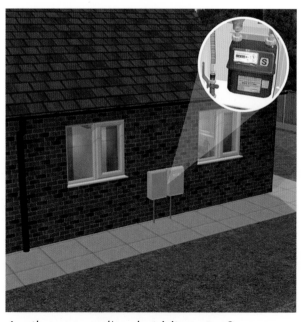

Are there gas and/or electricity meters?

Cables, pipes, flues and water supply

Where are the current cable, pipe and flue runs?
Is the hot water system on mains pressure or gravity?
If gravity fed, is the cold water tank well insulated and covered?

Other internal considerations include cables, pipes, flues and water supply

Insulation

What type of wall insulation does the building have?
What type of roof insulation does the building have?
What type of floor insulation does the building have?

Insulation: wall, roof, floor etc.

The hot water system

Does the building have a vented, unvented or thermal store hot water system?
Does the building have power showers?
Does the building have instantaneous water heating?

Hot water system

The heating system

Is the current heating system central or distributed?
What's the current heating system?
What's the current heat distribution system?
What are the controls for the current heating system?

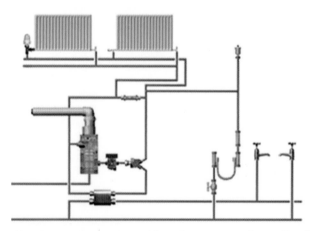

Heating system: central heating system, heat distribution system, controls etc.

Appliances and lighting

What appliances does the building currently contain?
What lighting does the building currently use?

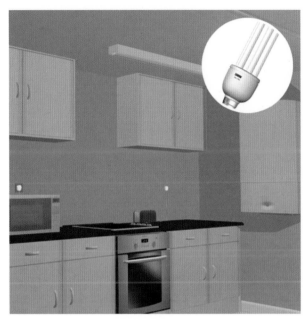

Current appliances used, lighting etc

ACTIVITY 26

What meters and distribution boxes should be observed during the internal site audit? Are there any other points to consider during an internal site audit?

The customer interview

The third stage of the selection process is to interview the customer with due care and consideration.

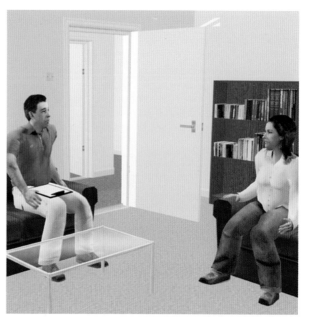

Interviewing the customer

Important questions to consider during a customer interview

Budget

What is the customer's budget for installation of renewable energies? What financial funding is available to them?

MONTHLY OUTGOINGS

Mortgage	£850
Council Tax	£120
Electric	£45
Gas	£40
Water	£40
Telephone	£15

SUPPORT

RHI Scheme

Feed in Tariff

Budgeting and financial funding

Occupancy

How many people live in the house?
What are the ages of the occupants?
Are the occupants out at work during the day?

Occupants of the household

Future growth plans

Is the customer planning any extensions to the house?
Is the occupancy likely to increase? (Customers might share this information, but it's polite not to ask directly.)

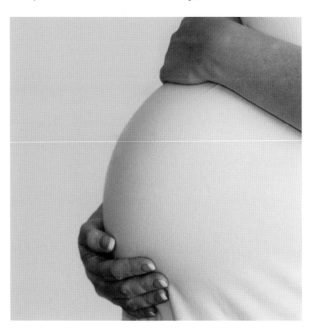

Plans for future growth: house extensions, increase in occupants etc.

Refurbishment plans

Does the customer plan any energy efficiency changes?
Does the customer have any existing renewable energy systems?
Is the customer concerned with the appearance of the building?

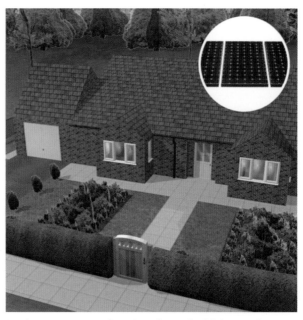

Is the customer planning for refurbishment?

Specific areas of interest

What are the customer's specific areas of interest?

Does the customer have any specific areas of interest?

Customer questions

Come well informed with case studies, photos and FAQs.

Insure you have all the information you need

ACTIVITY 27

How can you make a customer feel at ease during an inspection of the property and interview process?
Please list any points you think are helpful for communicating clearly with the customer about the interview and property inspection process.

SAVINGS AND COSTS

Introduction to the selection process

During the three stage external, internal and customer survey process,
you as the consultant obtain enough information to specify the various energy options for the property.

During the external phase, you can evaluate which technologies can be fitted to the property. During the internal phase, you can then audit which of these technologies will work with the building services currently used to supply the property with heating and electricity. Finally during the customer phase, the interview provides you with information as to the occupants' hopes and expectations for the future operation of their property.

The evaluation process enables you to specify energy options available

The three stages of the selection process

Once you have all the information you require on the building, you can begin the next phase of the selection process. In this phase you select the appropriate technologies. Again, this can be separated into three stages.

Stage 1

The first stage is to improve the energy efficiency of the building by recommending how the insulation, draught proofing, lighting, appliances and other features of the building can be improved. It is vital that the energy consumption of the property is minimized before new techniques for supplying the energy are employed.

First stage; improving the energy efficiency of the building

Stage 2

The second stage of the procedure is to improve the heating system of the building and this second stage has two parts. In part one, see if both solar water heating and a wood or pellet stove can be fitted. Both of these systems, if fitted, will provide part of the heating load. Remember all of the earlier advice from the modules when you are evaluating which technologies will work.

Second stage; improving the heating system of the building

In part two of the second stage, evaluate what is the best solution for the rest of the heating load. If the property is in a rural area, can a biomass boiler or GSHP be installed? If the property is in an urban area, can an ASHP or GSHP be installed? If the gas boiler needs replacing, can a micro CHP be fitted?

Typically, the greatest carbon savings are obtained with the biomass, then GSHP, then ASHP, then micro CHP systems. The costs associated with heating systems can often be offset with the Renewable Heat Incentive or similar grant or incentive scheme.

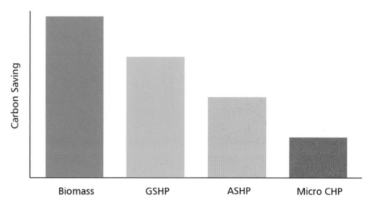

A bar chart illustrating the relationship between heating systems and carbon saving

Stage 3

In the third and final stage of the procedure, evaluate whether any electrical generation systems can be fitted to the property. Is the roof ideal for solar PV and can this PV system be fitted around or alongside the solar thermal system? Is the house in a windy rural location and, if so, is it suitable for a small wind turbine? Is there a stream with sufficient speed, flow and year round consistency to drive a hydro system?

The cost associated with these electrical systems can often be offset with a Feed-in-Tariff.

Third stage; evaluating if an electrical generation system can be fitted

In the above procedure, it is hoped that you can recommend several energy efficiency and renewable energy options for the customer to choose from along with some advice as to the benefits and drawbacks of each solution. You have probably spotted that micro CHP is, strictly speaking, an electrical gain rather than a heating benefit. However, the point of choice is a part of the heating solution options. Well done, you have now gone through a selection procedure for the property!

FUNCTIONAL SKILLS

Calculations in a practical context

ACTIVITY 28

Carbon Footprints

Work out the Carbon Footprint of a person living in Britain where the gas grid has an intensity of 0.205 $kgCO_2$/kWh and the electricity grid has a carbon intensity of 0.541 $kgCO_2$/kWh. The person shares a house with their partner and has no children. They use 18 000 kWh of gas for central heating and 4500 kWh of electricity. The person also drives a car which produces 150 $gmCO_2$/km and they drive 11 000 km/annum. Would the person's carbon dioxide output be higher or lower in Germany and Sweden respectively?

ACTIVITY 29

As well as energy for heating, electricity and private transport, what other factors should be considered to fully calculate this person's full carbon footprint? And considering these two final factors, how would the person minimize these two factors. Please don't include public buildings and workplaces but please explain why these should not be included and what you might do to minimize carbon output from these places.

Clue: think about your own lifestyle. Where do you consume energy beyond heat, electricity and private transport?

FUNCTIONAL SKILLS

Collect and interpret data

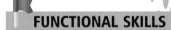

ACTIVITY 30

You are an energy consultant who advises householders on the Energy Efficiency and Renewable Energy improvements they can implement on their properties. Looking at the bungalow in the picture shown, you have noted the following information as you conducted the external evaluation, internal audit and customer interview sections of the survey:

- The customer says that they want to refurbish the building with lots of energy efficiency and renewable energy measures. They have come into a small inheritance so money, whilst an important consideration, is not a hindrance to the various energy options.

- The building is occupied by two adults and two children and has three bedrooms, one bathroom, kitchen, dining room, living room and hallway. It is located on the edge of London.

- The electrical system is modern and the house has spare slots on the consumer unit.

- The central heating and plumbing system has fairly old radiators for space heating, a 150 litre vented cylinder for HW provision, an old open galvanized cold water tank in the loft, TRVs on all the radiators and a modern room and cylinder stat with seven day programmer for central heating control.

- The insulation levels of the property are as follows:

  - Floor has 25 mm of insulation

  - Cavity walls have no visible signs of insulation being installed

  - The loft has 100 mm of insulation

  - The windows and doors are all single glazed

  - The draught proofing is poor

- The appliances, lighting and all electrical systems are all A or A+ rated.

- The building's design heat loss before insulation is 12.3 kW and if you implement the full range of energy efficiency measures, it will probably come down to about 8.1 kW.

- The front roof faces South-East and there are no trees or other objects to shade the roof. You have kept photos of the external building so that you have a good series of reminders when you are preparing the advice.

Please list in bullet form, the list of first, energy efficiency measures and then renewable energy measures that could be implemented on this property and any other thoughts that might be appropriate. If you want to relocate the building to another city or a rural area, than this is also good as it's useful to think about buildings in your own vicinity. The provided answers are based on a typical bungalow on the edge of London; however, these suggested answers would also often apply elsewhere on the edge of a city.

CHECK YOUR KNOWLEDGE

1. Which of the following is incorrect:

☐ a. Before considering any form of renewable energy solution in a building there is a three stage selection process that must be completed.

☐ b. This starts with an external evaluation of the building.

☐ c. It is followed up by an internal audit of the building.

☐ d. It concludes with a customer statement.

2. The external evaluation includes the following. Which one is incorrect?

☐ a. The building type

☐ b. Orientation

☐ c. The insulation of the building

☐ d. The construction of the building

☐ e. The land

3. The internal audit includes the following. Which one is incorrect?

☐ a. Metering

☐ b. The number of occupants

☐ c. Hot water system

☐ d. Heating

☐ e. Appliances and Lighting

4. The third stage of the selection process is to interview the customer with due care and consideration. What is not an appropriate question?

☐ a. What is the customer's budget for installation of renewable energies?

☐ b. How many people live in the house?

☐ c. What is the customer's annual income?

☐ d. Does the customer plan any energy efficiency changes?

5. Which one is incorrect? The three stages of the selection involve:

☐ a. Improving the energy efficiency

☐ b. Supplying the heat

☐ c. Generating electricity

☐ d. Changing the customer's behaviour

Chapter 6

END TEST

END TEST OBJECTIVES

This end test will check your knowledge on the information held within this workbook.

The Test

1. In kW, how much power does the Sun provide per m^2 at the equator? Enter a number into the box:

    ⬚

2. Wind energy is created when pressure differences between hot and cold air cause the air to move.

    ☐ a. True

    ☐ b. False

3. What process does biomass use to convert the Sun's energy into food for growth?

☐ a. Photostasis

☐ b. Photosynergy

☐ c. Photosynthesis

4. Which greenhouse gas mainly causes the average global temperature to rise?

☐ a. Carbon Dioxide

☐ b. Nitrogen

☐ c. Oxygen

5. By 2020, the UK must obtain a percentage of its energy from renewable sources. What is this legally binding target? Enter a number in the box:

%

6. In a solar thermal hot water system, vented or unvented layouts can be used but thermal stores are not possible.

☐ a. True

☐ b. False

7. Heat drawn from the air, the ground or water is already hot enough to be useful.

☐ a. True

☐ b. False

8. When wood is ground into sawdust and compressed it is known as:

☐ a. Chips

☐ b. Logs

☐ c. Pellets

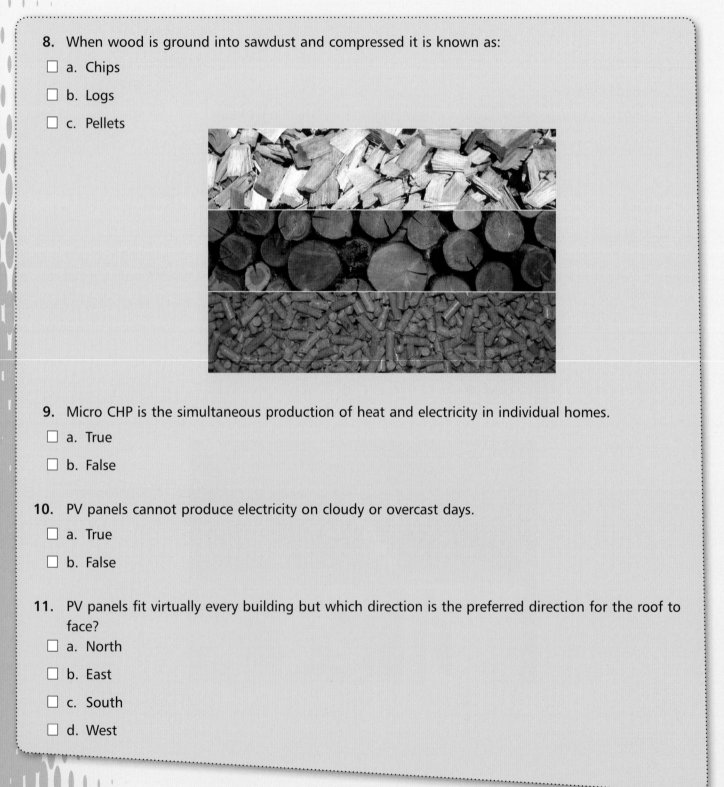

9. Micro CHP is the simultaneous production of heat and electricity in individual homes.

☐ a. True

☐ b. False

10. PV panels cannot produce electricity on cloudy or overcast days.

☐ a. True

☐ b. False

11. PV panels fit virtually every building but which direction is the preferred direction for the roof to face?

☐ a. North

☐ b. East

☐ c. South

☐ d. West

12. Do micro wind turbines need planning permission if the noise level exceeds 45 decibels at 1m from the neighbour's window?

☐ a. Yes

☐ b. No

13. What are the advantages and disadvantages of micro hydroelectricity?

☐ a. Expensive to install

☐ b. Have to be custom designed

☐ c. Minimal environmental impact

☐ d. Need a regular flow of water

☐ e. Quiet and reliable

☐ f. Generate a good income from FiT

Enter the correct letter in the following table:

Advantages	Disadvantages

14. Name two incentives to encourage conversion to renewable energy sources

☐ a. Feed in Incentive

☐ b. Feed-in-Tariff

☐ c. Renewable Heat Incentive

☐ d. Renewable Heat Tariff

15. The Merton Rule covers only residential and not commercial buildings

☐ a. True

☐ b. False

16. How many categories does the code for sustainable homes contain? Enter a number into the box:

[ ]

17. What is the order of the selection process for renewable energy solutions?

☐ a. Customer interview

☐ b. External evaluation

☐ c. Internal audit

Enter the correct letter in the correct box:

Step 1	Step 2	Step 3

18. Which checks are part of the external evaluation of a building, which are part of the internal audit and which are part of the customer interview?

☐ a. Appliances and lighting

☐ b. Building orientation

☐ c. Building type

☐ d. Budget

☐ e. Cables, pipes and flues

☐ f. Construction of the building

☐ g. Future growth plans

☐ h. The heating system

☐ i. The hot water system

☐ j. Insulation

☐ k. The land

☐ l. Metering

☐ m. Occupancy

☐ n. Outbuildings

☐ o. Refurbishment plans

☐ p. Specific areas of interest

Enter the correct letters in the table:

External evaluation	Internal audit	Customer interview

19. What is the order of the final selection of renewable energy technologies?

☐ a. Renewable electricity technologies

☐ b. Renewable and low carbon heating technologies

☐ c. Energy efficiency

Enter the correct letters in the table below:

Step 1	Step 2	Step 3

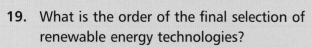

20. When choosing heating solutions, which two technologies are part of stage one, part-load heating solutions?

☐ a. Solar thermal

☐ b. Biomass

☐ c. Heat pumps

☐ d. Micro CHP

☐ e. Wood stoves

Answers

End Test Answers

1. The Sun provides 1 kW/m^2 at the equator

2. a – True.

3. c – Photosynthesis is the process which biomass uses.

4. a – Carbon dioxide is the main greenhouse gas.

5. 15 – By 2020 the UK must obtain 15 per cent of its energy from renewable sources.

6. b – Solar can be used with vented, unvented and thermal stores in both twin-coil and two cylinder layouts.

7. b – Low temperature heat from the environment must be upgraded to reach a useful temperature.

8. c – Sawdust is compressed into pellets

9. a – Micro CHP engines produce heat and electricity.

10. b – Panels can still produce electricity even on cloudy or overcast days.

11. c – South facing is the preferred direction.

12. a – Due to potential noise issues, micro wind turbines need planning permission if the noise level exceeds 45 decibels at 1m from the neighbour's window.

13. Advantages – c, e, f Disadvantages – a, b, d.

14. b, c – The two incentives are the Feed-in-Tariff and the Renewable Heat Incentive.

15. b – The Merton Rule covers residential as well as commercial buildings.

16. 9 – There are nine categories in the Code for Sustainable Homes.

17. Step 1 = b, Step 2 = c, Step 3 = a.

18. External evaluation = b, c, f, k, n Internal audit = a, e, h, i, j, l Customer interview = d, g, m, o, p.

19. Step 1 = c, Step 2 = b, Step 3 = a.

20. a, e – solar thermal and wood stoves are both part-load heating solutions.

Activity Answers

Activity 4

100 litres of water requires an 8 litre expansion vessel as the vessel needs to be twice the size of the expansion rate so that there is enough internal expansion vessel volume for the water to expand into. As water expands 4 per cent, 100 litres of water needs 2 * 4 litres = 8 litres of volume.

Activity 5

- Flat plates = 5 m^2 as you need 1 m^2 per occupant (four occupants) plus an extra 25 per cent because the roof is facing east.

- Evacuated tubes = 3.8 to 4.4 m^2 (this is rounded up from 3.75 to 4.375 m^2) as four occupants require 3 to 3.5 m^2 and this needs to be multiplied by 25 per cent.

- As the existing cylinder is vented, it is probably most effective to replace the existing cylinder with a 200 litre twin-coil vented cylinder (a 200 litre cylinder is preferred here as the recommended option is 50 litres of storage per occupant). However, any number of options of twin-coil, two cylinder, vented, unvented or thermal store layouts could be used. It's just normally most cost effective to keep the circuit similar to the existing design to avoid further complications.

- As this solar system is located in a temperate climate (England), it is better to use a pumped solar circuit.

Please note that different solar manufacturers and suppliers have their own recommended optimum solutions depending on the performance of their own equipment. These answers are based on some typical values rather than the specific needs of individual components.

Activity 6

At 45°C flow temperature, ASHP has an SPF of 3.0 and GSHP has an SPF of 3.7. SPF stands for Seasonal Performance Factor. It is a measure of the average efficiency of the heat pump over the heating season and is discussed in more detail in the Heat Pump eLearning course. At 80 to 100 W/m^2 room heat loss, the following oversize factors and pipe spacings apply:

- Domestic Fan Convector = 2.4
- Standard Radiator = 3.1
- Fan Coil Unit = 2.6
- Tiled screed underfloor heating can have a pipe spacing of 300mm or less
- Tiled aluminium panel underfloor heating can have a pipe spacing of 150mm or less.

Please note that the other types of underfloor are stated as unable to perform at the design parameters stated.

Activity 7

The first diagram is probably good as the straight pipes are well spaced out and the pipes look to be between 0.8m and 1.2m deep. 0.8m deep is recommended as the minimum depth for ground loops so that they are not disturbed by ploughing or other surface level digging and no more than 1.2m deep, as beyond this depth, shuttering is required whilst constructing the trench. With the second and third diagrams, the vertical and horizontal slinky looks to be well spaced. However, it is difficult to see the depth of the trench so it would be difficult to comment on this aspect of the installation. Indeed, a fully accurate check can only ever be made in person. The final photo probably has too much pipework installed and it is expected that with this much pipe installed, the pumping losses will outweigh the heat gained.

Activity 8

First of all, check the property for its energy efficiency. Draught-proofing, double glazing, loft, wall and floor insulation all help to minimize the heat loss from the property. The heat distribution system will probably need to be upgraded to either oversize radiators or underfloor heating. Please see the HE Guide for further advice. After the house has been insulated and the heating distribution system improved, then a quiet ASHP in the back garden in the centre of the property will probably be less than 42 decibels 1m from the neighbour's nearest window. A vertical loop GSHP can probably be fitted in the front garden and if there is access for a trenching machine, a horizontal loop GSHP can probably be fitted in the back garden. Only accurate on-site measurements and careful design will confirm these points. It looks to be an extensive change; however, the cost and environmental benefits will be significant.

Activity 9

Charcoal is made from wood (or other organic matter) and the process is called pyrolysis. Coke is made from coal. Both substances are made by heating the original material in an oxygen-free atmosphere.

Activity 10

Plant life, water, carbon dioxide, sunlight. Oxygen is a by-product of photosynthesis.

Activity 12

3000 litre pellet store, 9000 litre woodchip store. The log store would be bigger than the woodchip store. Wood is a great renewable fuel. However, the customer must be aware of the increase in fuel storage required.

Activity 13

Either a log stove or a pellet stove could be fitted. A wood boiler would typically replace the gas fired boiler and provide all the heating for the property so this is not suitable for this customer. A wood stove would provide an additional environmentally friendly and attractive solution. As it is an urban environment, it would be

necessary to check on the local environmental requirements and also the environmental output from whichever stove option was fitted. You would also need to confirm that the room had adequate ventilation to support the particular wood or pellet stove selected.

Activity 14

Reverend Robert Stirling invented the Stirling engine in 1816. It is classed as an external combustion engine. A fuel cell can be powered by hydrogen, natural gas, methanol and other fuels. There is much research and development into mCHP fuel cells and these might make a big impact on the building services industry. CHP and mCHP systems can be driven off a large range of different engines, each with their own advantages and disadvantages.

Activity 15

A competent heating engineer will be required to fit the gas side of the mCHP unit and connect it to the heat distribution system and a competent electrician will be required to connect up the electrical elements of the system. The power utility will need to be informed that an electrical generating system has been fitted to the property. This will apply in all countries and regions.

Activity 16

The main two other materials would be cadmium telluride, (CdT) and copper indium gallium selenide/sulphide (CIGS). Many materials including some organic cells demonstrate the PV effect so this list is fairly long and there are many correct answers.

Activity 17

A suitable DC to AC inverter. A basic inverter rather than a grid compatible is all that is required in this scenario. Or the occupants could purchase a DC television which would probably consume less power than the AC television and its associated inverter.

Activity 18

Windmills ground wheat and other grains. Wind turbines were also commonly used to pump water. There are also many other mechanical uses of this mechanical power.

Activity 19

There is eight times more energy in the 10 m/s wind as compared to wind at 5 m/s. The maths behind this answer:

$10 * 10 * 10 = 1000$

$5 * 5 * 5 = 125$

$1000/125 = 8$

Activity 20

The tube is called a penstock and a modern turbine is more efficient than a traditional water wheel and so generates more energy from a given volume and flow rate of water.

Activity 21

The first point to note is that electricity is widely used to provide heat. The main source of heating is natural gas; however, electricity, oil, coal, wood, lpg, solar and other heat sources all provide some of our current heating requirements. We will ignore these effects in the maths below and just use the data provided and simplify to obtain a base level result.

Looking at the maths, we know that travel at 20 per cent will not be the biggest emitter of carbon dioxide so we will not include this source any further. So looking at the other two sources, if we assume that the 49 per cent heat is all gas (a loose assumption but adequate for our simple model) and the 31 per cent is electricity, we get:

Heat = $49 * 0.205 = 10.045$

Electricity = $31 * 0.541 = 16.771$

To obtain a ratio: electricity / heat = 16.771 / 10.045 = 1.67

Please note that there are no units (e.g. $kgCO_2/kWh$) in the above maths as we have developed a ratio.

So in our simplified model, electricity accounts for 1.67 times more CO_2 emissions as compared to heat. However, in reality, this is just a paper exercise and we need to make sure our policy framework addresses all three factors, heat, electricity and transport, if we want to make a significant change in energy policy and so reduce carbon.

Activity 22

There is no simple answer to this dilemma and it is a question that often exercises policy makers. In the case of the RHI, there is an annual incentive payment for 20 years. The main advantage of measuring rather than estimating the heating consumption is that it provides a fair payment for a fair rate of usage. The main two ways of addressing wastage is either:

- to make sure that the incentive payment is less than the cost of the fuel
- or to cap the consumption payment.

All solutions to all different policy dilemmas have different pros and cons.

Activity 23

In the export power line out to the grid.

Activity 24

For Qualicert, any of the following: Ademe, Erec, Epia, Estif, Aebiom, Egec, Ehpa, Qualit'EnR, Ait, Ceetb, Ebc, Cres, Cape, Enea.

In the UK, for electricity it could be Napit or NICEIC (and some others) and for gas, Gas Safe, for oil, Oftec and for solid fuel, Hetas.

Activity 25

Solid and cavity wall constructions are in common use across all of Europe. Timber frame construction is not so common in the UK but is widely used in several European countries. Three common roof coverings are plain tiles, profiled or flat concrete tiles and slate. The choice of roofing is often determined by the local geology and cultural heritage of the region.

Activity 26

If they are fitted, the electricity, gas and water meters should all be checked. Also, the electrical consumer unit (also called distribution board) should be checked for its age and condition. For example, if this consumer unit still contains fuses rather than MCBs, it indicates that the house will probably need rewiring as well as having any renewable electricity technologies fitted. Modern heating systems are also moving towards heating manifolds, especially if underfloor heating is used. Whilst you were conducting the internal site audit, did you locate all the drain points on the central heating circuit and check the condition of all the electrical plug sockets? There is a lot of useful detail that can be acquired at all stages of the inspection process. You might have also come up with several or many other useful points during this listing and discussion process.

Activity 27

The more relaxed you are (to a point), the more relaxed a customer is. The customer likes to see a presentable auditor and they are also happy to see the auditor/interviewer wearing appropriate clothing to be able to inspect both the outside and inside of the property. Make sure you have the right tools, safety and access equipment to safely conduct the visit and spend a few minutes before starting the inspection explaining to the customer exactly what you are going to do, how long it will probably take and anything else you might require.

Activity 28

Answer:

Heating $= 0.205 * 18000/2 = 1845$ kgCO$_2$/annum

Electricity $= 0.541 * 4500/2 = 1217$ kgCO$_2$/annum

Private transport $= 150 * 11000 / 1000 = 1650$ kgCO$_2$/annum

Their Personal Total $= 4712$ kgCO$_2$/annum.

Germany, even though it has lots of wind power, has a lot of coal fired power stations, whilst Sweden has a lot of hydro and nuclear power. Therefore, this person's carbon footprint would be higher in Germany and lower in Sweden.

Activity 29

The two other important carbon dioxide factors are food and public transport (including flights). Food will have a low carbon footprint if the person uses a lot of locally grown vegetables and high if the person eats lots of prepared meals or food with lots of food miles. If the person uses local buses and trains, they will have a low public transport footprint. Flights and long distance trains will increase their CO_2 output.

Public buildings and workplaces would not be included to avoid national double accounting of carbon outputs. The idea is that people should be responsible for their own outputs and businesses and organizations responsible for their own outputs.

It would often be more difficult for the person to influence the carbon output from their workplaces and other public buildings. If they were the member of a club, church or other community group, they could make suggestions for minimizing energy consumption to the group. Workplace managers are normally pleased with ideas for reducing energy consumption in the workplace as this demonstrates employee engagement and also reduces energy bills.

There are many different carbon footprint calculators online, some very good and others with more mixed results.

Activity 30
Energy Efficiency

The external site inspection, internal audit and customer interview have all provided us with the information as listed above. This indicates that the:

- Electrical system is modern and energy efficient and so no further improvement is necessary
- The insulation levels could all be improved with the following:

    - Floors, it is difficult to insulate the floors as this will reduce the door height. However, another 25mm could be fitted under the existing flooring if the doors were also cut at the bottom

- Walls could be cavity wall insulated
- The loft could have another 200mm of insulation added
- The single glazed windows could be changed for Low-E argon double glazed units and the doors also brought up to a similar specification
- The house could be draught-proofed.

Renewable Energy

There is a huge range of RE options available for this building. Looking at the part-load heating options:

- A wood stove is probably a good option depending on how clean the stove burns the wood and the local smokeless zone requirements
- Either evacuated tubes or a flat plate solar thermal water heating system would work and as the roof faces south-east, it will obtain almost as much energy as a south facing roof. You would need to fit a new cylinder, probably a vented twin-coil of 200 litres (50 litres/occupant). You have also advised the customer that they must, for hygiene and bacteria protection purposes, change their old open cold water tank for a modern, plastic, insulated and screen filter ventilation cold water tank.

Looking at the full load heating options:

- An ASHP
- A vertical borehole GSHP
- A pellets boiler or stove with a back boiler
- Or a mCHP

would probably all work. The only option that looks not to be available is the horizontal loop

GSHP. However, after carefully surveying and calculation, this might also be an option if the collector was looped all the way around the building. Please note that it is unusual to have so many heating options available.

It is also important to note that either HP option will only work if a low temperature heat emitter system is also fitted and this could either be low temperature radiators or a low rise aluminium panel underfloor heating circuit.

The control system is modern with an up-to-date specification. The main next step to improve this would be to weather compensate the controller so that the full advantages of low temperature heating circuits were fully exploited.

Looking at the electrical options, this is far more limited. There is no stream nearby to run a hydro system and because this is in London, the wind speed is probably too low to justify a turbine. However, there is plenty of room on the roof for both a solar thermal and solar PV system to sit alongside each other.

Check your knowledge answers

CHAPTER 1

1. b – Fossil fuel is ultimately derived from the Sun's energy. It is stored solar energy from many millions of years.

2. Solar, Biomass, Wind, Wave – Currently, Biomass followed by Hydro and then Wind are the top three sources of renewable energy harvested on the planet.

3. d – It provides 16 kWh of heat at 2/3 of 24 kW in an hour.

4. b – Nuclear energy is the result of decaying atoms. Coal, oil and gas are fossil fuels.

5. c – Most climate change occurs naturally. Man is changing the climate beyond natural changes. IPCC stands for International Panel on Climate Change.

6. a – The UK has to achieve 15 per cent renewable energy by 2020.

7. b – Geothermal
   d – Tidal

8. b – Greenhouse gases in the atmosphere trap heat and raise the temperature of the atmosphere
   c – Millions are highly likely to be at risk of extreme weather events if measures are not taken against climate change.

CHAPTER 2

1. d – Northern Europe, especially in summer has a good supply of solar energy. Solar power can be used for space heating, industrial process heat etc. If evacuated tubes were much better than flat plates or vice versa, then this technology would be the only one on the market. Both tubes and plates work well when they are effectively employed.

2. a – Building and water regulations should and in certain cases, must be followed. A competent plumber can fit the plumbing but unless he or she has roofing competency, they cannot fit the solar collector. Solar also works with thermal stores. Passive systems use gravity.

3. c – Solar panels on a listed building or in a conservation area do not automatically qualify, building control ensures that buildings are safe, healthy, accessible and sustainable for current and future generations and solar pipework should always be insulated.

4. d – Heat pumps are classified as renewable energy, they need to be well designed, installed, commissioned and maintained and a heat pump has three major components.

5. c – With an open fire, 80 per cent to 90 per cent of the heat is lost up the chimney.

6. c – Most micro CHP units run on natural gas. However, CHP units can be run on a variety of fuels and as long as the fuel source makes economic sense, then an appropriate CHP unit will be available on the market.

7. Disadvantage, advantage, disadvantage, disadvantage, advantage, advantage.

8. b – Because when wood burns it only releases as much carbon dioxide as the tree has absorbed in its lifetime.

9. c – Micro Combined Heat and Power.

CHAPTER 3

1. d – A group of PV panels is known as an array. PV panels don't have in-built storage. Grid disconnected PV systems normally use a battery. PV comes in monocrystalline, multi-crystalline and amorphous layouts.

2. a – Some FiTs have a generation tariff and an export tariff. Others have just an export tariff. Different policies suit the needs of different countries. Inverters have several safety features as well as converting DC to AC. The BCC disconnects the PV panel from the battery when the battery is fully charged and disconnects the electrical load from the battery when the battery is fully discharged.

3. c – All wind turbines below 50 kW are classed as small, mini or micro. An exposed location rather than all of Northern Germany is good for producing wind power. To be effective, a wind turbine needs an average wind speed of no less than 5 m/s.

4. d – Small wind turbines work best in exposed locations without turbulence, they require planning permission for noise and appearance issues and vertical axis are not as efficient as horizontal axis wind turbines.

5. a – Very few locations are suitable for a micro hydroelectricity system.

CHAPTER 4

1. a – Fuel.

2. d – Grants are a form of financial incentive.

3. d – FiT stands for Feed in Tariff, RHI stands for Renewable Heat Incentive and the Code for Sustainable Homes is an environmental regulation that includes energy.

4. d – Solar Keymark certifies the Solar Collector not the installation. MCS stands for Microgeneration Certification Scheme and it provides greater protection (not financial returns) for the consumer.

5. a – 5 MW.

6. c – Competent Persons Scheme.

CHAPTER 5

1. d – It concludes with a customer interview.

2. c – Insulation is an internal audit issue.

3. b – The number of occupants is a customer interview issue.

4. d – The customer's annual income is not an appropriate question as energy rather than financial advice is being provided. Their available budget is a suitable question.

5. d – Changing the customer's behaviour. This is a building rather than behaviour change project. However, providing information, both verbal and written in a tactful manner on optimizing energy consumption is recommended.

Glossary

Anaerobic Digestion Is a series of processes in which micro-organisms break down biodegradable material in the absence of oxygen.

Airflow The movement of air through a space.

Amorphous In Solar PV panel terms, amorphous or thin film coatings do not have a regular crystalline structure and as the material is coated onto a surface, it would appear to have a haphazard structure under a microscope. Less light absorbing material is required to create a PV cell, hence these amorphous panels are often also called thin film. This often makes them the cheapest to produce but also leads to reduced efficiency of around 5 to 7 per cent. They have also become popular due to their flexibility, lighter weights, and ease of integration. Some more modern thin film panels come in layers which increase both efficiency and cost.

Bali Road Map After the 2007 United Nations Climate Change Conference on the island of Bali in Indonesia, the participating nations adopted the Bali Road Map as a two-year process to finalizing a binding agreement for 2009 in Copenhagen.

Biodegradable The chemical breakdown of materials by a physiological environment. The term is often used in relation to ecology, waste management and environmental remediation (bioremediation).

Biomass Biological material which is derived from living, or recently living organisms, such as wood, waste (hydrogen) gas and alcohol fuels. It is commonly plant matter grown to generate electricity or produce heat. Biomass is a renewable energy source.

Building Control Certificate A certificate that will be issued by an approved inspector for the compliance with Building Regulations in the proposed building work plans.

Carbon Dioxide (CO₂) A colourless, odourless gas that is formed during combustion, respiration and organic decomposition. Carbon dioxide emissions are considered to be a major cause of climate change.

Carbon Monoxide (CO) A colourless, odourless and tasteless gas that is very toxic to humans and animals.

Carbon Neutral Producing no carbon emissions, or balancing the amount of carbon released with an equivalent amount sequestered or offset.

Climate Change This is a change in the statistical distribution of weather over periods of time that range from decades to millions of years. Climate change may be limited to a specific region, or may occur across the whole Earth. Human activities such as burning fossil fuels which emit greenhouse gases, contribute to climate change. In the UK, 40 per cent of emissions are caused by individuals.

Central Heating A system that supplies space and hot water heating to a building from a single heat source through ducts or pipes.

Competent Persons Scheme (CPS) A Competent Person is a person who has been registered with an approved scheme by the Department for Communities and Local Government (DCLG). A Competent Person is adjudged to be sufficiently competent to self-certify that all work complies with the relevant part of the Building Regulations (England and Wales).

Compressor In reference to heat pumps, the electrical driven compressor increases the pressure and temperature of the refrigerant and pumps it around the circuit.

Condenser In reference to heat pumps, high temperature heat produced by the compressor is released in the condenser where the refrigerant changes from a gas to a liquid.

Convection Current In reference to airflow, convection currents are the movement of air caused by heated air rising and cooler air descending.

Copenhagen Accord The Copenhagen Accord is the document that delegates at the United Nations Climate Change Conference (UNCCC) agreed to 'take note of' at the final plenary session of the Conference on 18 December 2009.

Department of Energy and Climate Change (DECC)
This is a British government department created on 3 October 2008 to take over some of the functions of the Department for Business, Enterprise and Regulatory Reform (energy) and Department for Environment, Food and Rural Affairs (climate change).

Domestic Hot Water (DHW) Water which is heated and supplied for washing and bathing via taps or showers. DHW is not necessarily potable water and should be handled carefully to manage both Legionella and scalding risks.

Evaporator In reference to heat pumps, low temperature heat is gained in the evaporator where the liquid refrigerant is boiled and changes to a gas.

Expansion Valve In reference to heat pumps, refrigerant from the condenser passes through the expansion valve to lower its pressure ready for re-boiling from a liquid to a gas in the evaporator.

Export Tariff A scheme in which excess electricity generated and exported back to the grid is purchased.

Feed-in-Tariff (FiT) FiT or renewable energy payments is a policy mechanism designed to encourage the adoption of renewable energy sources and to help accelerate the move toward grid parity. Energy suppliers will make two types of payments to householders and communities. One payment will be to those who generate their own electricity (Generation Tariff) from renewable or low carbon sources. The other will be for the excessive electricity generated which has been exported back to the grid (Export Tariff).

Fossil Fuels These are fuels such as oil and coal that are formed by natural resources such as anaerobic decomposition of buried dead organisms. The age of the organisms and their resulting fossil fuels is typically millions of years, but can exceed two billion years. These fuels contain a high percentage of carbon and hydrocarbons.

Generation Tariff Payment made by energy suppliers to households and communities who generate their own electricity from renewable or low carbon sources.

General Permitted Development Order 2008 (GPDO)
Is an aspect of town and country planning which allows people to undertake minor development under a deemed grant of planning permission, therefore removing the need to submit a planning application. Permitted development is currently set out in the Town and Country Planning (General Permitted Development) Order 1995 (amended in 2008).

Geothermal Earth's temperature increases with depth, outward heat flows from a hot interior.

Greenhouse Effect This is a process by which reflected and infrared energy leaving a planetary surface is absorbed by some atmospheric gases, called greenhouse gases.

Greenhouse Gases Gases in an atmosphere that absorb and emit radiation within the thermal infrared range. This process is the fundamental cause of the greenhouse effect.

Ground Source Heat Pump (GSHP) A system that uses heat from the ground to heat (or cool) a building. It uses the Earth as a heat source in the winter or a heat sink in the summer.

Head In Micro Hydroelectricity terms, head is the term used to describe the difference in height between the water source and the water outflow.

Heat Exchanger A heat exchanger is a device designed to efficiently transfer heat from one medium to another.

Heat Transfer The movement of heat for one medium to another.

Hydroelectricity Electricity generated by hydropower, i.e., the production of electrical power through the use of the gravitational force of falling or flowing water. It is currently the most widely used form of renewable energy.

Inexhaustible In renewable energy source terms, inexhaustible sources are those which will not run out.

Intergovernmental Panel on Climate Change (IPCC)
IPCC is a scientific intergovernmental body tasked with evaluating the risk of climate change caused by human activity. The panel was established in 1988 by the World Meteorological Organization (WMO) and the United Nations Environment Programme (UNEP).

Kilowatt (KW) The kilowatt is equal to one thousand watts. This unit is used to express the output power of engines, the power consumption of tools and machines, the heating and cooling power used and any other forms of power.

Kilowatt-hour (KWh) The kilowatt hour is a unit of energy equal to 1000 watt hours or 3.6 megajoules. The unit of energy is how much power is consumed in a time period e.g. kilowatt hours.

Kyoto Protocol The Kyoto Protocol is a protocol to the United Nations Framework Convention on Climate Change (UNFCCC or FCCC), aimed at fighting global warming.

Low or Zero Carbon energy source (LZC) Low-carbon power comes from sources that produce fewer greenhouse gases than traditional means of power generation. It includes zero carbon power sources, such as wind power, solar power, geothermal power and (except for fuel preparation) nuclear power.

Megawatt (MW) The megawatt is equal to one million watts. Many events or machines produce or sustain the conversion of power on this scale.

Merton Rule The Merton Rule is named after the Merton Council, the first prescriptive planning policy that required new commercial buildings over 1000 square meters to generate at least 10 per cent of their energy needs using on-site renewable energy equipment.

Micro Combined Heat and Power (mCHP) Micro-CHP is an extension of the now well established idea of cogeneration (the use of a heat engine or a power station to simultaneously generate both electricity and useful heat). It is one of the most common forms of energy recycling to the single/multi family home or small office building.

MCS MCS is the first product and installer certification scheme to cover all the microgeneration technologies. MCS used to be called the Microgeneration Certification Scheme.

Monocrystalline In Solar PV panel terms, monocrystalline panels are the most expensive. A silicon ingot is extracted from a liquid vat of silicon and cools into a single crystal cylinder. The cylinder shaped ingots are then cut into 0.3mm thick slivers and arranged into PV panels. They are the most efficient however, at 15 per cent to 18 per cent efficiency.

Penstock A penstock is a sluice or gate or intake structure that controls water flow, or an enclosed pipe that delivers water to hydraulic turbines and sewerage systems.

Photosynthesis Using the energy from sunlight, it is the process that converts carbon dioxide into organic compounds, especially sugars. Photosynthesis occurs in plants, algae and many species of bacteria.

Photovoltaic (PV) The word photovoltaic comes from photo meaning light and voltaic which in turn means electricity. Solar Photovoltaic or PV cells convert the energy from the Sun directly into electricity.

Polycrystalline In Solar PV panel terms, poly or multicrystalline panels are less expensive to produce than monocrystalline panels but also less efficient at 14 per cent to 15 per cent efficiency. The silicon is either cut from cast square ingots or drawn in flat thin ribbons from molten silicon.

Refrigerant A refrigerant is a compound used in a heat cycle that reversibly undergoes a phase change from a gas to a liquid. Traditionally, fluorocarbons, especially chlorofluorocarbons were used as refrigerants, but they are being phased out because of their ozone depletion effects and more environmentally benign refrigerants have and are being introduced.

Renewable Heat Incentive (RHI) A scheme which provides financial support to those who install renewable heating technologies such as ASHP or GSHP, solar thermal and biomass boilers, etc. The scheme supports heating at all scales including domestic, commercial, industrial and public sectors.

Secondary Burning Systems Air combustion systems found in modern log burning stoves and other combustion processes which burn the volatile gasses driven from the wood fuel and significantly reduce the unburned carbon. Modern boilers often have primary air introduced before burning, secondary air into the combustion zone and tertiary air after the combustion process.

Solar Keymark The Solar Keymark is an official quality mark of the European Committee for Standardization (CEN), developed with ESTIF to overcome the varying testing and certification requirements in different EU countries. It accredits the quality of solar thermal collectors and systems.

Solar PV A Solar Photovoltaic System uses solar cells to convert light into electricity.

Solar Trade Association An association representing UK based solar companies. The association promotes the uptake of solar technologies to domestic, commercial and industrial applications.

Stirling Engine A Stirling engine is a heat engine that operates by cyclic compression and expansion of air or other gas, the *working fluid*, at different temperature levels such that there is a net conversion of heat energy to mechanical work.

Tailrace (Micro Hydroelectricity) In Micro Hydroelectricity terms, a tailrace is the channel of water flowing back to its natural watercourse.

Terminal A point or part that is at the end of a system. This can be an opening where ducting terminates, allowing air to flow in or out of a ventilation system.

Tertiary burning systems Air combustion systems found in modern log burning stoves which burn the volatile gasses driven from the wood fuel and significantly reduce the unburned carbon (see secondary burning systems).

Thermal Mass The capacity of a body or object to store heat. A building with a high thermal mass reduces temperature fluctuations within the building as it absorbs heat when the surroundings are hotter and slowly releases heat when the surroundings are cooler.

Tidal Tides are the rise and fall of sea levels caused by the combined effects of the gravitational forces exerted by the Moon and the Sun and the rotation of the Earth.

Turbine A turbine is a rotary engine that extracts energy from a fluid flow and converts it into useful work. Please note that a fluid can be a gas or a liquid.

Ventilation The removal of air containing moisture and pollutants from within a building, replacing it with fresh air from outside.

Watts The watt is a derived unit of power in the International System of Units (SI). The unit measures the rate of energy conversion. It is defined as one joule per second.

Zero-Carbon Producing no carbon emissions.

Index